ISBN 978-1-332-06996-5
PIBN 10280025

This book is a reproduction of an important historical work. Forgotten Books uses state-of-the-art technology to digitally reconstruct the work, preserving the original format whilst repairing imperfections present in the aged copy. In rare cases, an imperfection in the original, such as a blemish or missing page, may be replicated in our edition. We do, however, repair the vast majority of imperfections successfully; any imperfections that remain are intentionally left to preserve the state of such historical works.

1 MONTH OF
FREE
READING

at
www.ForgottenBooks.com

By purchasing this book you are eligible for one month membership to ForgottenBooks.com, giving you unlimited access to our entire collection of over 1,000,000 titles via our web site and mobile apps.

To claim your free month visit:

www.forgottenbooks.com/free280025

English
Français
Deutsche
Italiano
Español
Português

www.forgottenbooks.com

Mythology Photography **Fiction**
Fishing Christianity **Art** Cooking
Essays Buddhism Freemasonry
Medicine **Biology** Music **Ancient**
Egypt Evolution Carpentry Physics
Dance Geology **Mathematics** Fitness
Shakespeare **Folklore** Yoga Marketing
Confidence Immortality Biographies
Poetry **Psychology** Witchcraft
Electronics Chemistry History **Law**
Accounting **Philosophy** Anthropology
Alchemy Drama Quantum Mechanics
Atheism Sexual Health **Ancient History**
Entrepreneurship Languages Sport
Paleontology Needlework Islam
Metaphysics Investment Archaeology
Parenting Statistics Criminology
Motivational

STUDIES OF HEREDITY
IN RABBITS, RATS, AND MICE

BY

W. E. CASTLE

RESEARCH ASSOCIATE OF THE CARNEGIE INSTITUTION OF WASHINGTON

PUBLISHED BY THE CARNEGIE INSTITUTION OF WASHINGTON
WASHINGTON, 1919

CARNEGIE INSTITUTION OF WASHINGTON
PUBLICATION No. 288

[From the Laboratory of Genetics of the Bussey Institution]

JUL 24

PRESS OF J. B. LIPPINCOTT COMPANY
PHILADELPHIA

CONTENTS.

PART I. FURTHER EXPERIMENTS UPON THE MODIFIABILITY OF THE HOODED CHARACTER IN RATS.

PAGE

Selected races crossed with wild... 2

PART II. THE INHERITANCE OF WHITE-SPOTTING IN RABBITS, WITH SPECIAL REFERENCE TO GAMETIC CONTAMINATION.

Dutch 5
English 19
Relation of Dutch to English... 25

PART III. OBSERVATIONS ON THE OCCURRENCE OF LINKAGE IN RATS AND MICE.

Tables 37
Bibliography.. 56

STUDIES OF HEREDITY IN RABBITS, RATS, AND MICE.

By W. E. CASTLE.

I. FURTHER EXPERIMENTS UPON THE MODIFIABILITY OF THE HOODED CHARACTER OF RATS.

In publications Nos. 195 and 241 of the Carnegie Institution of Washington reports have been made on the results of a series of experiments designed to show to what extent a mendelizing character, the hooded pattern of piebald rats, may be altered by selection or by crossing. At the last report (Castle and Wright, 1916) the plus-selection series had been carried without out-crossing through 16 successive generations, in the course of which the mean grade of the offspring had advanced from +2.05 to +4.13, in terms of the arbitrary grading scale depicted in both previous publications. Since then the plus-selection series has been carried through four additional generations of selection (17 to 20) and the mean of the selected race has been raised to 4.61. In some respects this part of the series is less satisfactory than that previously reported on, because smaller numbers of animals were available from which to select and the selection has therefore been less rigorous. The race has unmistakably fallen off in vigor and fecundity in later generations. It is uncertain whether this should be ascribed to inbreeding alone, uncorrected by selection for vigor (as in Miss King's experiments), or to increase in the prevalence of disease, or to both causes. Certain it is, however, that notwithstanding increasing care in regard to feeding and sanitation a very large proportion of our breeding-pens in the case of the selected races produce no young at all.

Tables 1 to 4 show in detail the grade distribution of the young produced by plus-selected parents of generations 17 to 20. The numbers of young produced in each of these generations are respectively 351, 420, 280, and 92. The mean grade of the young advances from 4.13 (in generation 16) to 4.48 (in generation 17), remains practically stationary in generations 18 and 19 at 4.46 and 4.49 respectively, and then advances again (in generation 20) to 4.61.

The minus-selection series—which at the last report had been carried through 17 generations, with an advance in mean grade of the young from grade minus 1.00 to grade minus 2.70—has now been carried through 4 additional generations of selection. (See tables 5 to 8.) But this race has shown even poorer vitality than the plus-selection race, so that in order to keep it alive practically no young could be rejected as parents and consequently no further progress has been made. The numbers of young recorded for the four additional generations have been 330, 130, 79, and 35 respectively, and their mean grades −2.84, −2.89, −2.78, and −2.74. The race is now practically stationary in grade, but seems likely soon to become extinct despite our strongest efforts to keep it alive. Notwithstanding the fact that the race is verging on extinction after long-continued close breeding, the variability of the hooded character is still as great as ever. The standard deviation ranges from 0.25 to 0.45 within about the same limits as in the previous 17 generations from minus-selected parents. In the plus-selection series the standard deviation was also fully as high in the last generations as it had been in the previous 10 generations. Only the initial 7 generations had shown an appreciably higher variability.

SELECTED RACES CROSSED WITH WILD.

We may now inquire what happens to the races modified by selection in opposite directions, when they are crossed with an unselected, non-hooded race, the wild race. This question was considered in some detail, so far as the plus-selected race is concerned, in a previous publication (1916), where it was shown that a cross of the plus-selected race reduced the grade of the hooded character, undoing in a measure the work of selection. Selected animals which mated with their like should have produced young of about mean grade 3.75, actually produced hooded grandchildren, extracted in F_2 from a wild cross, of mean grade 3.17, a falling off in grade of over 0.50. A second cross with wild showed no further falling off, but instead a movement in the reverse direction to 3.34 (a movement probably not significant, in the light of further experiments). A third cross with wild has been made on a small scale; 19 hooded grandchildren extracted in F_2 from this third cross have a mean grade of 3.04. (See table 9.) It seems probable, therefore, in the case of the plus-selected hooded character, that the maximum effect exerted by the residual heredity of the wild race is to reduce the hooded character in grade by about three-fourths of a grade. Selection in 10 previous generations had elevated the grade of the hooded character by about $1\frac{3}{4}$ grades. A cross with wild eliminated less than half of this change. The remaining change must be ascribed to changes effected in the course of selection.

The minus-selected hooded race has also been crossed with this same wild race. Originally of grade −1, it had been altered by 15 generations of selection to the extent of about $1\frac{1}{2}$ grades, to mean grade −2.54. Females of generations 15 to 16 (table 10) and of grade −2.75 were crossed with wild males of the same race used in the crosses of the plus-selected race. The extracted hooded F₂ young were highly variable, ranging in grade from −2.25 to +3.00, mean −0.38, a remarkable change in the plus direction of over 2 grades.

A second cross with the wild race (table 11) brought about a further movement of the mean in a plus direction but by a somewhat smaller amount, 1.39 grades, the mean of the twice-extracted hooded young being +1.01. A third cross with the wild race (table 12) brought still further contamination of the hooded character, which now ceased to vary below grade +1.00, and had a mean of +2.55 in the case of over 100 thrice-extracted hooded young, this being a change in the plus direction of 1½ grades. It will be observed that the hooded grand-children of ♂2068, table 12, the most *plus* in character of the hooded grandparents, are very similar in grade to the hooded young resulting from three crosses of the *plus*-selected hooded character with the same wild race (table 9). In other words, the same wild race, when its residual heredity is made fully effective by repeated crosses, brings both the plus-selected and the minus-selected hooded lines to a pheno-type of common grade. This shows, contrary to my earlier opinion, that what has really happened in the case of the selected races was more largely due to residual heredity than to any change in the gene for the hooded character itself. My critics have been wrong when they insisted that selection could not change racial characters that mendelize and change them permanently, and when on this ground they denied to gradual change through selection an important part in the evolution of characters and thus of races. But my critics have been right when they insisted that evidence is wanting that change in *single genes* occurs other than spontaneously, uninfluenced by sys-tematic selection.

II. THE INHERITANCE OF WHITE-SPOTTING IN RABBITS, WITH SPECIAL REFERENCE TO GAMETIC CONTAMINATION.[1]

One of the commonest color variations of mammals is white-spotting—the occurrence of wholly unpigmented areas in the skin and the hair arising from it. Small unpigmented areas are frequently found in the coats of wild mammals, as, for example, in the fur of wild mice, rats, the common North American rabbit (*Lepus sylvaticus*), and cavies (*e.g., Cavia cutleri*). The white-spotting found in these wild forms is usually not extensive. It consists of a white "star" in the forehead or a spot on the chest, or at the end of the tail, or on a foot. Such locations of the white-spotting suggest a deficiency of pigment in the skin, either where it closes together in the median line during development of the embryo or at the extreme limits of its peripheral extension during development. At places where the skin regenerates after injury, even in self-colored animals, a white spot frequent'y develops. This is especially noticeable on the backs and shoulders of horses where the harness has "galled" them. That such slight congenital deficiencies of pigment as occur in wild mammals are hereditary has been shown by Little in the case of the house-mouse, and by Phillips and myself (unpublished observations) in the case of the field-mouse, *Peromyscus*. We observed in a colony of *Peromyscus* reared from animals taken in Massachusetts the occurrence of individuals having tails partly or wholly white. This condition was found to be a Mendelian recessive character in crosses. After one or two selections of white-tailed individuals, we noted extension of the white area on to the belly.

In some wild mammals the white-spotting is more extensive, taking the form of a definite pattern, as in skunks, the harp-seal, and the Malay tapir. The color pattern of skunks, while characteristic, is known to vary slightly, the value of a pelt increasing with the amount of black which it contains, a fact which the incipient industry of skunk-farming in the United States notes with interest. Selective breeding is being directed toward the establishment of all-black strains and no doubt it will ultimately be successful.

White-spotting is so common in the domestic animals as to need no comment. White-spotting in more or less definite patterns characterizes the majority of our breeds of cattle, horses, dogs, and swine. Often the pattern is so definite and so strongly inherited as to constitute a sort of trade-mark of breed purity, as in Hereford (white-faced) cattle, Dutch belted cattle, and Dalmatian coach-dogs. Much

[1] Valuable assistance in the conduct of this investigation was given by my former pupil, Prof. H. D. Fish.

interest attaches to the inheritance of these patterns in crosses, a subject which has been studied for some years at the Bussey Institution. The present paper will deal with the subject of white-spotting in domestic rabbits.

Patterns of white-spotting mendelize without known exception but with some irregularity as regards dominance. In some cases white-spotting is not expressed in the heterozygote, but if expressed at all in the heterozygote its expression is always stronger yet in the homozygote. When the character is nearly or quite suppressed in the heterozygote, we may call it recessive; when the character is strongly expressed in the heterozygote, we may call it dominant. But neither term is applicable without qualification in the way that recessive is applicable to complete albinism in rodents.

With this qualification of terms, it may be said that there occur among domestic rabbits two forms of white-spotting, probably of independent origin and certainly of quite different genetic behavior, since one is recessive and the other dominant in crosses with the same race of unspotted rabbits. The dominant form of white-spotting is found in the so-called English rabbit and its inheritance has been discussed by Castle and Hadley (1915). The recessive form of white-spotting is found in Dutch rabbits, as observed independently by Hurst and by Castle. Punnett assents to this conclusion with the qualification that the inheritance is possibly not that of a simple (one-factor) sort.

DUTCH.

In September 1910 three standard-bred Dutch rabbits were obtained from a fancier who had bred and exhibited prize-winning animals derived from stock imported from England. They resembled grades 7, 8, and 9 respectively (plate 1). The female proved to be sterile and was ultimately discarded. One of the males ($\male$3037, grade 7) was mated with two heterozygous English does, which produced self-colored blacks as recessives when mated with English bucks of their own race. For the present we shall consider only the non-English young produced by these matings. Such young would be of the same character as those produced by self-colored animals mated with Dutch, since they would arise from a *self* gamete furnished by the mother and this would be fertilized by a *Dutch* gamete furnished by the father. Six young of this character were produced of the accompanying grades.

Grade.	No.
1	2
2	1
3	3

(See *table* 13).

This same male was mated also with 3 Himalayan albino does of a race entirely free from spotting but which lacked the color factor. Potentially these does were self-colored. This cross produced 18 young of the accompanying grades.

Grade.	No.
[1]0	3
1	13
2	2

(See *table* 13).

[1] Grade 0 signifies a self animal, *i.e.*, one *without* white spotting.

Two does derived from the first-mentioned cross and one derived from the second were employed in various matings presently to be described. The results observed in the case of all three were so similar that they may conveniently be described together. It will be borne in mind that all are F_1 hybrids between Dutch and self.

When crossed back with the other original pure Dutch buck ($\male$3036, grade 9), these three does produced 20 young of the grades shown in table 14. We get here indications of segregation into two groups, one like the F_1 mothers in grade, the other like the Dutch father, but no sharp line of division separates the two.

The same three F_1 females were also mated successively with an F_1 male from each of the two crosses already described, with the results shown in table 15. The results are similar in both cases, but it will be noticed that the lower grade F_1 male (5029, derived from the Himalayan cross) produced F_2 young of slightly lower grade. The F_2 range extends from 0 to grade 5 inclusive, average 1.80. The back-cross range was from 1 to 7 inclusive, average 4.60.

Certain of the F_2 young and the back-cross young of grade 4 or higher, which presumably would be homozygous for the Dutch character, if it mendelizes, were employed in building up a race of Dutch rabbits for further study. This was done by back-crossing the selected does a second time with the pure buck, $\male$3036, grade 9, with the results shown in tables 16 and 17. Young were obtained which ranged from grade 1 to grade 17, but which grouped themselves round two modes situated at about grade 6 and grade 15 respectively. We shall presently consider the distribution further.

These same does were also mated with a male similar in origin to themselves, viz, $\male$5167 (table 18), a typical and evenly marked Dutch buck of grade 7, produced by the original back-cross (table 14). He bred in all respects like his father ($\male$3036, grade 9) when mated with the same does, producing a bimodal group of Dutch young of only slightly lower mean grade than the young which his father sired, as might be expected from the fact that his grade was less than his father's grade. (See table 18.)

It was now evident that we had secured a race of Dutch rabbits which produced only Dutch young and which derived their Dutch character exclusively from the two bucks 3036 and 3037, and yet which fluctuated in grade around two different modes. In fact, it was soon discovered that the two original Dutch bucks were themselves heterozygotes of two different types of Dutch pattern which corresponded with the two modal conditions found among their descendants. Our next task was to isolate these two types in homozygous form. This was easy in the case of the higher grade (whiter) type, which proved to be recessive. A male of this "white" type, 6175, grade 17, when mated with females of the same sort produced only one type of Dutch young varying around grade 16. (See table 19.)

By studying the results of various matings of our Dutch does it was found possible to classify them in three categories: (1) The "white" type of grade 15 to 17, which produced only the "white" type when mated with bucks of the same sort, as already described. (See plate 2, fig. 19.) (2) A "dark" type of grade 1 to 7, which when mated with bucks of the white type produced no "white" offspring, but only those of the dark type. These mothers were evidently *homozygous "dark,"* the other pure type (see plate 2, fig. 20). (3) The third type of doe is scarcely distinguishable from the pure dark type except by breeding test. It consists of heterozygotes between the two types, a little whiter on the average than the dark type, but not conspicuously so. When they are mated with "white" bucks, young of two types are produced in about equal numbers, viz, heterozygous darks and pure whites. The original Dutch bucks from which the entire race was derived were both of this heterozygous dark type.

Neither the dark type nor the white type isolated from this race of rabbits conforms closely with the ideal Dutch type of the fancier (our grade 8). The one is usually too dark and the other too white. It seems probable that the fanciers in breeding "prize-winners" have consciously or unconsciously been producing heterozygotes, very much as in the case of the Andalusian fowl. Certain it is that all the rabbits which we have produced from this stock, which would have any chance of winning a prize at an exhibition, have been heterozygotes between these two types.

While the experiments with standard-bred fanciers' Dutch rabbits were in progress, and after the white and the dark types of Dutch had been isolated, a third type of Dutch was discovered which kept cropping out in a stock of black-and-tan rabbits under observation for another purpose. This stock was derived from a single pure-bred black-and-tan buck which had been crossed with various other stocks of rabbits then in the laboratory. The Dutch pattern had been introduced as a recessive character in a certain yellow rabbit of unknown pedigree obtained by purchase. When the descendants of this yellow rabbit were bred with each other, certain of the young produced were Dutch marked. This type of Dutch resembled the fanciers' type of Dutch (grade 8, plate 1) so far as the head markings were concerned, but the belt was very narrow and placed far forward over the shoulders. (See plate 2, fig. 21.) Because of its origin within the black-and-tan stock we have adopted the name "tan" Dutch to distinguish it from the other two types.

In describing its variations we have used the same set of grades (shown in plate 1) which were used in classifying the variations of the other two types, but it must be understood that rabbits of the dark and of the tan type which are given the same grade are not exactly alike in pattern. The grade is assigned to express roughly the total amount of white-spotting on the animal. A rabbit of grade 3 in the tan series will usually have a whiter head with a broader white spot on the nose but with a narrower collar than one of the same grade in the dark series. With this explanation it may be stated that tan Dutch rabbits bred with each other have produced 40 Dutch young with the grade distribution shown in table 21.

The variation is close about grades 3 and 4, the race being very uniform in character for a white-spotted race. In origin it is derived from a single gamete which introduced the character as a recessive into the original yellow ancestor.

Having now secured 3 distinct strains of Dutch rabbits, it was our next task to determine what were their genetic relationships to each other, whether they were allelomorphs or due to wholly independent factors; whether due to single or to multiple genetic factors, and whether these factors were constant or variable. Before these questions can be intelligently discussed the variability of each type by itself must first be known. That of the tan race has just been referred to. It is shown in table 21. It will be observed that the mean grade of the young has a tendency to rise with the grade of the parents.

The variation of the uncrossed "dark" race is shown in table 20. The same homozygous buck (6701, plate 2, fig. 20) was mated with ten different homozygous dark does ranging in grade from 2 to 5. They produced 172 young ranging in grade from 1 to 7, mean 3.30. The variation is shown graphically in text-figure 1, D. The higher-grade mothers, it will be observed, produce higher-grade young, although the differences are not striking.

The variation of the uncrossed "white" race is shown in table 19 and text-figure 1, w. Two bucks of grade 17 were mated with 8 does of grade 15, 16, or 17; they produced 59 young with the same range of variation as the mothers and of the mean grade 16.25. Again we observe a tendency for the higher-grade mothers to produce the higher-grade young.

The same homozygous white buck (6175) which was used in matings recorded in table 19 was mated also with 5 homozygous does of the dark race, with the results shown in table 22 and text-figure 1, F_1, D × W. These matings produced 28 young of mean grade 7.28. All the young, from their parentage, should be heterozygotes between white and dark Dutch, like the original animals from which these two races were isolated. In reality they agree closely with the foundation stock in grade.

Table 23 shows the results of matings of homozygous white bucks (one of which was the same individual, 6175, as sired the young of tables 19 and 22) with dark does which were heterozygous for white and so were like the foundation stock. The young thus produced numbered 130 and fall into two distinct groups, each varying about a different mode. The two groups apparently do not overlap. Taking the minimal class, grade 12, as on the line between them, there are 65 individuals below this class and 64 above it. The former should be

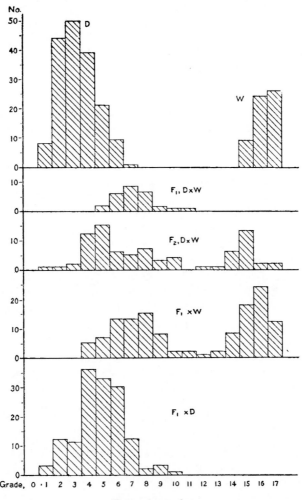

TEXT-FIGURE 1.

heterozygotes like the young recorded in table 22. Their mean grade is very similar, 7.04 as against 7.28. The latter are evidently homozygous white like the young of table 19. They have a similar but slightly lower mean grade, viz, 15.56 as against 16.25. This result indicates a 1 : 1 segregation of white and dark Dutch and that accordingly white and dark are allelomorphs. If so, an F_2 generation should show a 3 : 1 segregation. To test this point we may summarize from tables 16 to 18 all matings between two heterozygous parents (those which produce both dark and white young). In this way we get the totals shown in table 28, "white × dark." Two groups of young are here shown, one dark (grade 10 or lower), the other white (grade 12 or higher). The numbers of individuals in these groups are 56 and 25 respectively (a rather poor 3 : 1 ratio); their mean grades are 5.82 and 14.40. (See text-figure 1, F_2, D×W.) The dark group evidently includes homozygotes (mode on 5) and heterozygotes (mode on 8) which overlap in the intervening region. Accordingly all facts thus far noted indicate that white and dark are allelomorphic forms of Dutch marking.

One other test of this hypothesis is possible. F_1 may be back-crossed with the dark race. The result of such a test is shown in table 29, "F_1 (white × dark) × dark," and text-figure 1, $F_1 × D$. The expectation here is the formation of two groups of equal size, homozygous and heterozygous dark, with modes on 3 and 7 respectively. In reality the intervening grades are the modal ones. This does not disprove segregation. The flat-topped variation curve observed is exactly what we might expect from the combination of two simple variation curves which overlap. Compare F_2, S×W, text-figure 2.

The several facts developed as regards the relation of "dark" to "white" Dutch pattern are presented graphically in text-figure 1. The variability of the uncrossed races is shown at the top; the races are distinct, monomodal, and do not overlap in range. Immediately below is shown the character of F_1. It also is monomodal and it is intermediate. Next lower is shown the variability of F_2. The extracted white race is here seen to have a lower mode than the uncrossed white race and the extracted dark race has a higher chief mode than the uncrossed dark race. In other words, the extracted races reappear in F_2 in character mutually modified, although their extreme range is the same. The mode of the heterozygotes lies between the modes of the homozygotes, as in F_1 just above.

The results of back-crosses with each of the parental races are shown in the lower part of text-figure 1. In the back-cross with the white race ($F_1 × W$), the 1 : 1 segregation is unmistakable; in the back-cross with the dark race ($F_1 × D$), segregation is obscured by the closeness to each other of the modes for heterozygous and homozygous dark Dutch, which results in producing a composite flat-topped curve. Everything indicates that dark and white are simple allelomorphs, but are quantitatively fluctuating and mutually modify each other in crosses. If they were not allelomorphs, individuals which contained neither form of Dutch should appear in F_2. None such is produced.

We may next consider whether white Dutch is a simple allelomorph of self or is composite. The pertinent facts are recorded in tables 27 to 29 and are shown graphically in text-figure 2. The same white male (6175) that was crossed with dark females was also crossed with self-colored (unspotted) females. (See table 27.) There were produced 33 young, all showing a small amount of white-spotting, ranging

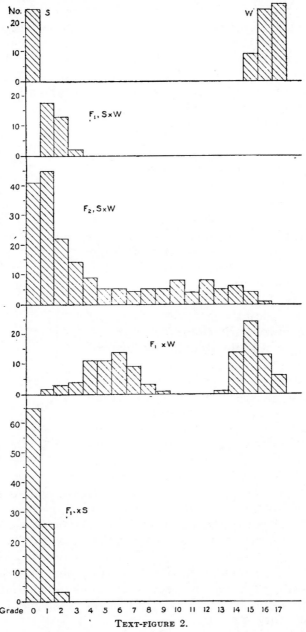

TEXT-FIGURE 2.

in grade from 1 to 3, mean 1.51 (see text figure 2, F_1, $S \times W$). F_2 is somewhat puzzling in character. (See table 28 and text-figure 2.) Nearly 25 per cent of the 191 young recorded are extracted selfs, but if we take the whitest 48 individuals to be extracted "whites" it is evident that they are considerably modified from the condition of the uncrossed race, since their mode lies at 12 to 14, not at 16 to 17, and the highest grade (17) of the uncrossed white race does not reappear at all. The back-cross of F_1 with white must be investigated before one can interpret the F_2 result with confidence. But the back-cross (table 29 and text-figure 2, $F_1 \times W$) makes it very clear that segregation occurs on a 1 : 1 basis. The 116 young thus produced fall into two groups which do not overlap and each of which is monomodal.

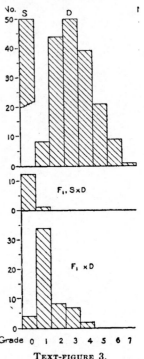

TEXT-FIGURE 3.

Each contains 58 individuals. But the extracted groups show mutual modification. The mode of the lower group is not at grade 1, as in F_1, but at grades 4 to 6, while the mode of the upper group is not at 16 to 17, as in the uncrossed white race, but at 15.

The back-cross of F_1 with self (table 29 and text-figure 2, $F_1 \times S$) gives a variation nearly covering the combined range of self and F_1, as expected, but the two expected groups (if they are distinct) lie so close together that it is impossible to separate them.

The several facts developed in crosses of white with self indicate that white is a simple allelomorph of self. If so, and if dark is also an allelomorph of white, as indicated by text-figure 1, then self and dark should be allelomorphs of each other. Such is probably the case, but crosses of self with dark are inconclusive because the two conditions are so close to each other on the grading scale that it is difficult to demonstrate segregation. See tables 27 and 29 and text-figure 3. This figure shows (at the top) the variation of uncrossed dark in relation to self. Below is shown the variation of F_1, self being usually dominant. A back-cross with dark ($F_1 \times D$) produces less than the expected proportion of self (nearly 50 per cent as indicated by F_1) and produces extracted darks of lower mean grade than the uncrossed darks. This is evidence of mutual modification of self and dark in the heterozygote, so that they emerge in the gametes modified. So far, then, the evidence indicates that self, dark, and white are all allelomorphs, but that they fluctuate quantitatively and mutually modify each other when associated in heterozygotes.

We may now consider the relation of these three to tan Dutch. If tan is an allelomorph of either of the other forms of Dutch, it should be an allelomorph of all three conditions, in fact a *fourth* allelomorph.

To test this matter, crosses of tan have been made with each of the other three, white, dark, and self. Tan was found to be allelomorphic with self in the black-and-tan race where it appeared (table 12), but self being a purely negative term (meaning as here used unspotted) one could not be sure in advance that self, as the allelomorph of different kinds of Dutch, would be one and the same thing in all cases. This would require demonstration. It is necessary, then, to ascertain first whether tan Dutch is an allelomorph of white and of dark.

Crosses of tan with white have given the results shown in tables 26, 28, and 29, and also graphically in text-figure 4. F_1 is intermediate. F_2 is likewise intermediate, but varies to or into the range of the uncrossed races with indications of segregation of modified tan and modified white individuals. The absence of selfs in F_2 shows white Dutch and tan Dutch to be allelomorphs. The back-cross with white ($F_1 \times W$) produces individuals varying (as expected) all the way from the F_1 to the uncrossed white race, but with an apparent tendency to form modes on 10 and 15.

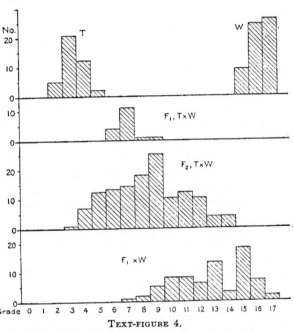

TEXT-FIGURE 4.

These, it seems, represent the mutually modified modal conditions of the F_1 and the white race. The extracted whites have their mode lowered to 15, instead of on 16 or 17, as in the uncrossed white race; and the extracted F_1's have their mode advanced from grade 7 (in the original F_1's) to grade 10.

The races *dark* and *tan* are similar to each other in grade (see text-figure 5), but differ in the location of their unpigmented areas, as already explained. Dark has a wider collar and a darker head; tan has a narrower collar and a whiter head. F_1 is nearly or completely self-colored, since *dark* tends to make the *head* pigmented and *tan* to make the collar pigmented. Consequently there is little space left unpigmented. F_2 is quite variable, some individuals being darker than either uncrossed race, while others are whiter. The range is from grade 0 to grade 11. (Text-fig. 5, F_2, D×T.) Indications of segregation are plainly seen in F_2, but it is impossible to be sure of the number of factors involved because dark and tan are so similar in grade. But we know that dark is an allelomorph of white. If it is also an allelomorph of tan, then F_1 individuals should produce dark gametes and tan gametes in equal numbers. These, in a cross of F_1 with white, uniting with white should produce two kinds of heterozygotes: (a) white-dark heterozygotes like those of text-figure 1, F_1, and (b) white-tan heterozygotes like those of text-figure 4. The former range from grade 5 to grade 11, the latter from grade 6 to grade 9. The ranges are similar and the mode of each group is on grade 7. Mutual modification of dark and tan would tend to extend the range of segregates in both directions, as observed in F_2 (text-figure 5).

The observed back-cross generation (table 29 and text-figure 5, $F_1 \times W$) is distinctly bimodal, as the hypothesis just formulated (that dark and tan are allelomorphs) would demand. Indeed, the evidence of 1 : 1 segregation are clearer than we should expect, but the modes of the expected groups are farther apart than we should have expected. (See text-figure 5.) The lower mode is at 4, not at 7 as in the F_1 produced by dark crossed with white; and the upper mode is at 12 to 14, not at 7, as in the F_1 of tan crossed with white. It appears, therefore, that while dark segregates from tan in the gametes formed by F_1 individuals, each segregates in an altered form, the dark having become darker and the tan lighter as a result of their association in the heterozygous F_1. This is indicated both by the bimodal condition and increased range shown by the cross of F_1 with white, as just described, and also by the wide range of the F_2 from the cross of dark with tan.

An alternative hypothesis, which has been given careful consideration and which in fact the cross of F_1 with white was especially designed to test, is this: that dark and tan are due to independent factors and that the two together constitute white. With this hypothesis the following facts harmonize: Dark crossed with tan produces in F_1 and F_2 individuals which are self-colored (i.e., which have no white-spotting) as well as others which are whiter than either the dark or the tan race. The former may be interpreted as those which lack (or are heterozygous for) both dark and tan; the latter as those which have both dark and tan. But it should be observed that none of the 275 F_2 young which have been produced extend into the range of the uncrossed white race, where about 25 per cent of them should lie if the hypothesis is correct. Compare text-figures 1 and 5.

Another way of testing the hypothesis that dark and tan are due to independent Mendelian factors is the cross of F_1 with white. If the hypothesis is correct, F_1 individuals should produce four kinds of gametes, viz, those which will transmit (a) both dark and tan, (b) dark alone, (c) tan alone, and (d) neither dark nor tan. By the hypothesis white is dark plus tan. Therefore, white crossed with F_1 should produce four kinds of zygotes: (a) dark and tan united with white (dark and tan), equivalent to homozygous white; (b) dark united

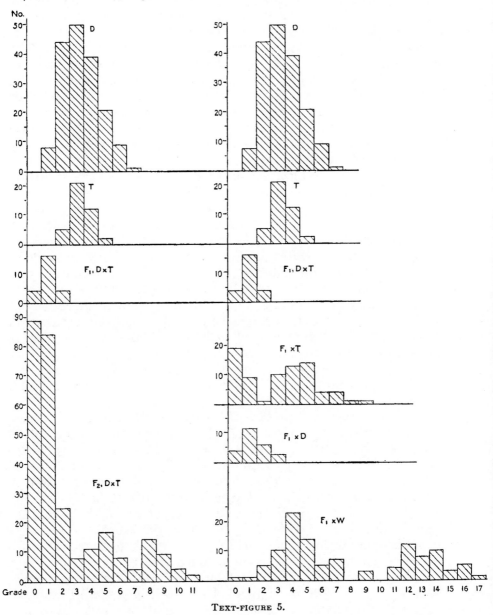

TEXT-FIGURE 5.

with white; (c) tan united with white; and (d) self (neither dark nor tan) united with white. The observed modal values of these four kinds of combinations are respectively, in terms of grades: (a) 16 (text-figure 1, W); (b) 7 (text-figure 1, F_1); (c) 7 (text-figure 4, F_1); and (d) 1 or 2 (text-figure 2, F_1). That is, this cross should, under the hypothesis considered, produce a trimodal figure, with a mode for 25 per cent of the individuals at either end of the grading scale and with a still larger mode (for 50 per cent of all individuals) at an intermediate point. But what is actually observed is very different. The figure (text-figure 5, $F_1 \times W$) is not trimodal but bimodal, and neither of the modes is where the hypothesis demands that modes should be. This is conclusive evidence against the correctness of the hypothesis in question, but is entirely in harmony with the alternative one, that dark and tan are allelomorphs but segregate in modified form, one on the whole darker, the other on the whole lighter than before they were crossed with each other.

This matter of modification on crossing is one deserving further consideration. It is in evidence in all the crosses made. It is clearest where the races crossed differ most in grade, and it seems to consist in a partial obliteration of those differences. Thus the uncrossed white race has its mode at 17, the self race at 0. (See text-figure 2.) In F_2 the extracted whites contain no individual as high in grade as 17, and the highest mode lies at 12 to 14, facts which indicate that white has been lowered in grade by its association with self in the F_1 zygotes. The back-cross of F_1 with white also shows modification but intermediate in amount, as might be expected from the fact that, in the case of each zygote formed, only one of the two conjugating gametes had been subjected previously to modifying influences, viz, that one which was furnished by the F_1 parent. The mode in this case lies at 15, instead of at 17, as in the uncrossed race, or at 12 to 14, as in the F_2 extracted whites.

That the extracted white has less influence in whitening an F_1 zygote than the uncrossed white is shown further by a comparison of F_1 self $\times$ white (text-figure 2) with the back-cross of F_1 with self. F_1 had its mode at grade 1 and ranged upward to grade 3, but contained no self individuals. If extracted white were identical with uncrossed white in its whitening influence, then in the back-cross with self half the zygotes should be of grade 1 or higher. But in the observed back-cross less than one-third of the zygotes show any white, viz, 29 out of 94, and these are lower in grade than the original F_1's, viz, 1.10 as compared with 1.51.

The self character also emerges modified after the cross with white. For the F_1 zygotes, consisting of pure self united with pure white (text-figure 2, F_1, S $\times$ W) were all close to self in grade, with a mode on grade 1 and ranging upward only to grade 3. But the zygotes formed by *extracted self* united with pure white produced in the back-cross of F_1 with white (text-figure 2, $F_1 \times W$) range in grade from 1 to 9 with a broad low mode at 4 to 7. Evidently they have been much increased in grade in the direction of white.

Again, in the crosses of dark with white (text-figure 1), we see the mutual modification of the contrasted conditions taking place. Uncrossed dark has its mode on grade 3, and uncrossed white on grade 17. Extracted white as seen in F_2 has its mode on grade 15, but in the back-cross with pure white it has its mode on the intermediate grade, 16.

As regards the modification of dark, it will be observed that the original F_1 individuals are of lower average grade than the lower group of individuals produced by the back-cross with white. The mode of the former also is on grade 7, that of the latter is on grade 8. The one consists of pure dark united with pure white, the other of extracted dark united with pure white. The extracted dark has evidently been *whitened*, exactly as extracted white has been *darkened*.

Uncrossed tan and uncrossed white have their modes on 3 and 17, respectively (text-figure 4). The mode of F_1 is on grade 7, but the mode of extracted tan united with white is on grade 10, as seen in the back-cross of F_1 with white. This shows that extracted tan has been whitened as compared with uncrossed tan. That white has had its grade lowered by the cross with tan is also shown in the back-cross of F_1 with white. Its mode lies at 15, not at 17 as in uncrossed white.

If, in the several cases considered, crossing tends to mutual modification and assimilation to each other of the contrasted conditions brought together in the cross, why does crossing of dark with tan extend rather than shorten the range of variation, producing in F_2 individuals darker than either uncrossed race and others lighter than either uncrossed race? The answer to this question is perhaps to be found in the imperfection of our scale of grades. The scale is a linear one, whereas the variation is not entirely linear; for dark Dutch and tan Dutch have the same modal grade, 3, yet are different in somatic character, as has already been stated. Tan Dutch has a white head and narrow collar, dark Dutch has a dark head and wider collar. In F_1, if pigment simply dominates whiteness, we may expect to get a dark head and a narrow collar simultaneously, *i. e.*, a condition with less white than either parent possessed, which in general is the result observed.

Since uncrossed dark Dutch varies down to grade 1 and uncrossed tan down to grade 2, an extension of the dark areas due to crossing naturally carries the pigmentation in F_1 down to grade 0 (self) in a certain percentage of cases (1 in 6 observed). See text-figure 5. In F_2 the percentage of selfs is still larger, being about 1 in 3. There are also found in F_2 whiter individuals than either uncrossed race contained. How these have arisen is indicated by the back-crosses of F_1 with dark and with tan respectively. The back-cross with dark (text-figure 5, $F_1 \times D$) produces a monomodal group closely resembling the F_1 group, but with slightly higher range. This shows that the potent factor in lowering the F_1 range was the *dark* gamete, since it is the only common factor entering into both crosses. On the other hand, the back-cross with tan ($F_1 \times T$) shows bimodal variation with the modes at 0 and 5 respectively. This indicates that the gametes formed by F_1 are really of two types, extracted dark and extracted tan. The former uniting with pure tan produces a group like F_1 but apparently of even slightly lower grade, since the modal condition is now 0, not 1. The extracted tan gametes uniting with pure tan produce a group like pure tan, but of apparently *higher* grade, since the mode is now on 5, not 3 as in pure tan. Hence extracted tan is potentially of higher grade than pure tan, a conclusion supported by F_2 from dark crossed with tan, for here we observe that extracted tan meeting extracted tan produces zygotes of grade 5 to 8 or even higher, whereas pure tan does not exceed grade 5. Yet, to return to the imperfection of our grading scale, these higher grades consist merely in combining a wider collar with the same form of head markings as are found in the tan series of grade 4 or 5. Hence it appears that tan is regularly modified through its contact with dark in an F_1 zygote in the way of acquiring a *wider collar*, whereas dark is modified by the same agency in the way of acquiring a *narrower collar*. Yet there is no indication that head marking and collar marking are due to distinct single genetic factors, but merely that they are qualitatively different. This difference tends to disappear through mutual influence in the heterozygous condition, but the difference disappears more rapidly in *collar* markings than in head markings; hence the extended range of grades in F_2. Because their *collars become more alike* the extracted darks rank lower in grade and the extracted tans rank higher; for it will be recalled that the uncrossed darks, though having wider collars than the tans, were graded lower on account of their dark heads. The collar changes, then, are actually blending in this cross, as in all the others studied, but give the appearance of segregation with differences emphasized, merely because of the inadequacy of our linear grading scale to record simultaneously changes in head and in collar markings when these occur with unequal rapidity.

It has already been shown that we have conclusive evidence that F_1, from the cross of tan with dark (text-figure 5), produces *two* types of gametes, not *four* types, this evidence being (a) the bimodal variation seen in the back-cross of F_1 with tan and (b) the bimodal variation seen in the cross of F_1 with white. These results indicate that tan and dark are to be regarded as allelomorphic but mutually modifying conditions, as had already been found to be true for tan and white, for dark and white, and for self and white. We have, then, a condition of multiple allelomorphs in white-spotting patterns of Dutch-marked rabbits, which includes the forms self, dark Dutch, tan Dutch, white Dutch, and possibly many other types or conditions of white-spotting which with sufficiently accurate observation might be distinguished from each other.

The foregoing observations show unmistakably that the several members of this allelomorphic series tend, as a result of crosses, to become more like each other. This has been described as mutual modification, but it should be expressly stated that in the light of our experiments with rats "modification" need not be regarded as change in the nature of a single gene, but merely as equalization of the residual heredity additional to the single genes which produce monohybrid ratios.

ENGLISH.

In November 1909 there were received at the Bussey Institution four "English" rabbits, 1 male and 3 females, bred by R. W. Wills, of Hornerstown, New Jersey. In terms of the grading scale shown in plate 3, the male was of grade $2\frac{1}{4}$; the females were of grades 2, $2\frac{3}{4}$, and 3, respectively.

In matings of the male with each of the 3 females, there were produced both English and self-colored young, as shown in table 30; of the former, 21; of the latter, 8. The self young were later found to produce no English young when bred *inter se*. Hence it seems clear that English is a dominant Mendelian character, that self is recessive in relation to it, and that the 4 English parents were all heterozygous dominants.

The question now arose whether homozygous English rabbits could be produced and why English rabbits were not regularly bred in homozygous form. We did not have long to wait for an answer to these questions. Table 30 shows that the English young of our 4 original English rabbits fall into two groups quite different in appearance. Of the 15 young which were graded, 5 were of grade 1 or $1\frac{1}{4}$, while 10 were similar to the parents in grade, varying from grade 2 to 3. The group of low-grade English was found to consist of homozygous individuals which produced only English young in crosses with each other or with selfs. The higher-grade group, twice as numerous in individuals, was found to consist of heterozygotes. These are preferred by the fancier because of their much more striking color-pattern. The homozygote is in appearance only an impure white animal, but the heterozygote is beautifully mottled. It is therefore clear why the fancier breeds heterozygotes. (See plate 3.)

Our original English buck, 2545, was also mated with self-colored does of several different sorts, viz, gray, cream, yellow, sooty yellow (tortoise), black, and black-and-tan. In regard to color inheritance, these matings gave us such results as are already familiar through earlier publications by Punnett, Hurst, and ourselves. We may therefore confine our attention to the behavior of the English pattern in crosses. Table 30, B, shows the results obtained; 26 English and 18 self young were recorded from these matings. No grade was recorded for 17 of the English young. The others varied in grade from $1\frac{1}{2}$ to 4, the mean being 2.80.   Evidently the heterozygous English produced by these matings with unrelated does were much more variable than those produced by matings of the original English individuals with each other.

It now occurred to us to see to what extent this variability could be carried farther in a plus direction by selection.   Accordingly we chose the grade 4 individual produced by ♀1492 as the starting-point of the experiment.   This individual had been recorded as ♂2711. He constitutes generation 1 of the selection experiment now to be described, all animals produced in those experiments deriving their English from him.   He was mated with a black-and-tan doe, with three black does, and six Himalayan (albino) does, all free from racial white spotting (English or Dutch).   These matings produced 56 young (table 31, B) equally divided between heterozygous English and self.   The English young were of higher grade than the English young produced by ♂2545.   (Compare tables 30 and 31.)   They ranged from grade 3 to grade 5, mean 3.89, as compared with a mean of 2.80 for the young of ♂2545, produced in similar matings.

Male 2711 was later mated to 5 of his heterozygous English daughters produced in the matings already described, and also to one of the resulting grand-daughters.   The character of their young is shown in table 31, A.   The mothers form generations $1\frac{1}{2}$ and $2\frac{1}{2}$ of the selected English race.   They vary in grade from $3\frac{1}{2}$ to 5.   They produced three classes of young: low-grade English (homozygotes), high-grade English (heterozygotes), and selfs.   Their respective numbers were 12, 11, and 8.   The mean of the low-grade English group was 1.52, that of the high-grade group was 3.93, which agrees closely with the grade of the same group produced by matings with non-English (self) does (table 31, B).

Grouping the mothers by grade, the relation shown herewith is observed between grade of mother and grade of young.   This indicates that selection of higher-grade parents would probably result in producing higher grade young.

Mother.	Mean of young.
3.50	3.83
4.75	3.80
5.00	4.25

A son of ♂2711, viz, ♂5086, generation $1\frac{1}{2}$, grade $4\frac{1}{2}$, was chosen to succeed him in the selection experiment.   This buck was mated with 10 different heterozygous English does, 5 of which had also been mated with his father.   (See table 32.)   There resulted 20 homozygous English young, 44 heterozygotes, and 35 selfs.   The mean grade of the homozygotes was 1.38, that of the heterozygotes was 3.96, averages not very different from those which characterized the young of ♂2711 (table 31).   Consequently no advance can be claimed as a result of the selection of ♂5086.   He was mated also with 3 black-and-tan does (table 32, B), producing thus 10 English young of mean grade 3.15, a lower average than that given by ♂2711 in matings with self does, but it must be borne in mind that the mothers were not all identical in the two cases. Black-and-tan does seemed in general to give lower-grade offspring than does of the self black and Himalayan races, with which ♂2711 had been mated.

We next used as sire in the selection experiment ♂5375, generation $2\frac{1}{2}$, grade $4\frac{1}{2}$, a son of ♂5086 by his half sister. (See table 33.) He was mated with 9 different heterozygous English does, all but one of which had also been mated with his father. They produced English young of somewhat lower mean grade than they had borne by the father, ♂5086. (See tables 32 and 33.) 17 homozygous English young were of mean grade 1.20; 29 heterozygous English young were of mean grade 3.79; there were also 25 self young. Again the higher-grade mothers produced the higher-grade young. Hence there was evidently material favorable for selection among the mothers, if not among the fathers. This male was now discarded and replaced by an own brother of slightly higher grade, viz, ♂5555, generation $2\frac{1}{2}$, grade $4\frac{3}{4}$. (See table 34.)

This male (♂5555) was bred more extensively than any of his predecessors and produced higher-grade offspring. He was mated to the same does as his father and grandfather and also to a number of new ones which now became available. By all classes of does he produced higher-grade young than had any of his predecessors. He also produced higher-grade young by his mates of higher grade than by his lower-grade mates. (Table 34.) In his case, then, a second advance had been made by selection in the male line and the necessary variation was evidently present to make possible similar advances by selection in the female line. Male 5555 produced 41 homozygous English young, 125 heterozygous English, 2 ungraded English, and 65 selfs, or all together 168 English and 65 selfs.

The next sire tested was a son of ♂5555; viz, ♂6370, grade 5, generation 3, as regards selected ancestry. (See table 35.) He was more advanced in grade and generations than any male thus far tested and produced higher-grade young by females of the same grade. Many of the older females had now been discarded, but enough remained to form a standard of comparison between the genetic properties of this male and those of his predecessors. The mean grade of the heterozygous English young of this male was 4.66; the grade of his homozygous English young was 1.79. The corresponding figures for his father were 4.40 and 1.36 respectively.

Three other sons of ♂5555 were also tested by matings with substantially the same group of does, although tests in the case of the less promising ones were terminated sooner. Male 6420 (table 36) was of slightly lower grade than his father and was found to be probably inferior to him and so was soon discarded. His heterozygous English young were of mean grade 4.33.

Male 6071 (table 37), although of higher grade, gave no better results. His own brother, ♂6072 (table 38), born in the same litter and graded the same, did much better. He was bred very extensively and gave a record very similar to that of his father, ♂6370, who was of the same grade but had half a generation less of selected ancestry. Male 6072 had 75 homozygous English young, of mean grade 1.97 (father's record 1.79); he also had 159 heterozygous English young of mean grade 4.63 (father's record, 4.66). Increase in the grade of the young with increase in the mother's grade is very clearly shown among the young of this sire. (See table 38.)

The next male tested was 6964 (table 39) a son of ♂6071. He was discarded after a set of matings which showed him probably not better than his uncle, 6072, who was still in service. He had 21 homozygous young of mean grade 1.49 (his uncle's record was 1.97) and 29 heterozygous English young of mean grade 4.68 (his uncle's record being 4.63). Next was tested ♂7699 (table 40), son of ♂6072, who shared with his half brother, ♂6370, the position of best sire so far. All were of the same grade, 5. This male was mated with all available does and produced 354 recorded young. He has a better record than any sire so far tested. By heterozygous does he has sired 75 homozygous young of mean grade 2.31 and 149 heterozygous young of mean grade 4.80.

Another male of the same grade and generation as the foregoing, indeed his half-brother, being also a son of 6072, was tested, but appeared not to be superior to 7699 and so was soon discarded. This animal, 9532 (table 41), sired 16 homozygous English young of mean grade 2.53 and also 27 heterozygous English young of mean grade 4.73.

Three sons of the superior male, 7699, have since been tested, viz, 9806, 1212, and 534 (tables 42–44). The first one shows no probable superiority over his father, but the last two are more promising, each having produced a total of over 60 heterozygous English young with a mean close to grade 5. In the case of their father the corresponding group of young were of grade 4.80. The homozygous young produced by their father were of mean grade 2.31; those produced by the sons were of mean grade 2.87 and 2.95 respectively. Accordingly, as regards both heterozygous and homozygous young, the sons have a distinctly better record. This, no doubt, was due in part to the fact that their mates were of higher grade or from more highly selected stock, but it was not wholly due to this cause, for their half-brother (9806, table 42) did not show the superiority which they showed, even when mated with females of high grade and advanced generations. Hence we must conclude that these two males, 1212 (table 43) and 534 (table 44), were genetically superior to their father.

Table 45 shows the grade distribution of the young produced by a homozygous English male, 1173 (plate 3, fig. 6), when mated with does of the three categories used in testing heterozygous English males. It will be noted that he produced only English young, however mated, conclusive evidence of his homozygous dominant character. His heterozygous English young were of mean grade 4.77 and 4.84 by homozygous English and self does respectively. This male was a son of male 9532, table 41, with whose genetic character his own was very similar, judging by the grade of their heterozygous English young. His line was not continued.

Table 46 enables one to survey at a glance the summarized results of this entire selection experiment. The course of the experiment is followed only in the male line, because only in the case of the males is the number of young large enough to show beyond question the genetic properties of the individual. From the beginning of the experiment fluctuation was observed in the grade of the young produced, and this fluctuation was in part at least genotypic, since the higher-grade mothers have given higher-grade young in matings with the same male. That the fluctuation was also in part phenotypic is shown by a comparison of the records made by different males of the same grade when mated with the same group of females (tables 31–45).

The entire selection race derives its English character from ♂2711. This animal was a heterozygote deriving the English character in a single gamete from ♂2545, who was also heterozygous. Hence the English character had evidently changed in transmission from father to son, a sufficient refutation of the idea of unit-character constancy. Whether the change resulted from a directly changed unit-factor (gene) or from the introduction of one or more modifying factors is a matter for further consideration.

The advances made in the male line seem to occur as five successive steps corresponding roughly with generations of selected ancestry (table 46). The first advance comes with the selection of the (single gamete) male 2711, founder of the race; the next in the selection of his grandson, 5555; the third in two sons of 5555, viz, 6370 and 6072; the fourth occurs in the selection of two sons of 6072, viz, 7699 and 9532; the fifth is seen in 1212 and 534, sons of 7699. The direct line of advance is through 2711, 5555, 6072, and 7699. The *amount* of advance at each step, as indicated by the average grade of the young of these males, is shown in table 47. The rate of advance has evidently decreased as the experiment progressed.

That modification of the English pattern resulted immediately from the cross with self individuals of an unrelated race is conclusively shown in table 30. The original English male, 2545, produced by English mates heterozygous English young of mean grade 2.32; by self mates he produced heterozygous English young of mean grade 2.80, practically half a grade higher. By most of such mates the young were more than a grade in advance of those produced by English mates. Since this is the direct effect upon the English character of *one* dose of the self race, it might be supposed that *two* doses would have a greater effect, so that if it were possible to lift the English character bodily out of the English race and surround it with the complete residual heredity of the self race, an effect perhaps twice as great as that actually observed in the cross might be expected. Accordingly an advance of between 2 and $2\frac{1}{2}$ grades may be attributed to the residual heredity of the self race. Theoretically, if this residuum consisted of a number of independent factors, then full effect would be secured upon breeding with each other the highest-grade individuals of the cross-bred race, repeating this process generation after generation until each factor was present in a homozygous state. This is substantially the procedure which has been followed in the 5 full generations over which the experiment has extended. The advance realized amounts to about $2\frac{3}{4}$ grades.

Allowing for the fact that one of our arbitrary "grades" may not have the same genetic value as another, it seems probable that we have secured something more than the effect of the residual heredity of the self race employed in the original cross. This may have resulted either from a process of elimination from the heredity complex of factors which tended to lower the grade of the English character or from change in the heredity complex by some other process than addition or subtraction of factors—for example, by change in factors.

The important fact which this experiment demonstrates is the same as that shown in the selection experiment with rats, that the single characters which serve to identify our domestic races of animals and which give value to them, even though they conform with every criterion of unifactorial Mendelian heredity in transmission, do nevertheless vary through minute gradations. By reason of the fact that the residual heredity affects such characters, a cross into an unrelated race can not be made, except with the possibility, or usually with the probability, that the character or characters in question will be thereby modified. This fact was formerly expressed in the statement that "contamination" of unit-characters frequently follows upon cross-breeding—a form of statement, however, which was challenged by those who maintained that the gametes were "pure." Subsequent investigation has shown beyond question not only that unit-characters are frequently greatly modified by crosses, but also that they can be modified by selection alone unattended by crossing.

Those who formerly maintained the doctrine of gametic purity now shifted their ground, and while admitting that unit-characters might change, insisted that single factors or "genes" could not change. This is the doctrine of pure genes which Morgan has made so familiar. This doctrine it is difficult either to prove or disprove. Pragmatically speaking, it is of small consequence, since it is admitted (1) that single factors do sometimes change, leading to the formation of multiple allelomorphs; (2) that the action of single factors is not limited to any particular part of the organism, but may affect parts apparently unrelated; (3) that the total number of factors concerned in the genesis of even the simplest organisms must be very great; and (4) that in what should theoretically be "pure lines" (asexually reproducing organisms, Jennings) genetic changes are constantly occurring. Pragmatically, then, genetic variability by minute gradations is a reality, precisely as Darwin assumed it to be, and this fact allows races to be altered steadily and permanently by selection, either natural or artificial, as Darwin also assumed was the case. The hypothesis that stable organic forms come into being only suddenly, by abrupt changes from preëxisting forms and not by gradual modification—this hypothesis, the "mutation theory" as commonly understood, is not substantiated.

RELATION OF DUTCH TO ENGLISH.

It remains to consider the genetic relations to each other of Dutch and English spotting. Dutch, as we have seen, behaves as a recessive in crosses with self-pigmented races; English, on the other hand, behaves as a dominant. Dutch marking appears to result from a simple deficiency of pigmentation, as if in development the pigment supply failed at an extremity or at an embryological point of finishing-off. When Dutch marking is reduced to its lowest point of expression by selection or crossing, the only white visible is found at the tip of the nose, or on the toes of a fore-foot, or as a spot in the middle of the forehead. English spotting, on the contrary, appears to result from some positive inhibiting force, some agency which uses up the pigment-forming materials here and there in the epidermis and converts them into an end-product not colored but white. That English individuals possess all the agencies necessary for full pigment formation is shown by the fact that English parents may produce fully pigmented (self) young as recessives, which then produce only self young if mated with each other.

We have seen that there occur different forms of Dutch spotting, which apparently behave as allelomorphs, but which tend to become less distinct, one from another, when they are associated in the same zygote. It would seem probable that they represent quantitatively different stages of reduction in the amount of some substance carried in the germ-cell. But undoubtedly this substance, whatever it is, is located in the chromatin, since the defect is transmitted equally through egg and sperm. There are also qualitative differences among the different forms of Dutch, as for example between "dark" and "tan" Dutch, in one of which the white collar is more in evidence, while in the other it is the head markings that are more in evidence. Probably, then, the different forms of Dutch are variants of a single "locus," in the terminology of the chromosome theory. But the physiological and genetic behavior of English are so different that it would seem improbable that they are variants of the same gene.

Nevertheless, when English is crossed with Dutch, the two appear to be either allelomorphs or closely linked, as the following results show: Heterozygous English rabbits of grade 5 were crossed with "white" Dutch of grades 15 to 17. Two matings were also made of the homozygous English buck, 1173 (table 33, and plate 3, fig. 6), whose English character was of equivalent potency with that of heterozygotes of grade 5. These matings produced 26 English and 9 non-English (Dutch) young, which were graded (with no great exactness) as follows:

There was probably no real discontinuity in the grouping of the English young, but owing to the rough manner of grading them the numbers heap up on the even grades, fractional grades being neglected. The Dutch young are similar in grade to the young produced by crossing "white" Dutch with self. (Compare table 27.) But the English young are much whiter than heterozygotes between English and self. The latter are 4.75 to 5.00

Dutch young.		English young.	
Grade.	No.	Grade.	No.
0	3	$1\frac{3}{4}$	2
1	2	2	15
2	3	$2\frac{1}{4}$	1
3	1	$2\frac{1}{2}$	2
..	..	$2\frac{3}{4}$	1
..	..	3	5
Total	9		26

in grade when produced by the same English sires. (Compare table 45.) But the English-Dutch heterozygotes in no case are of higher grade than grade 3 and in the great majority of cases are no darker than grade 2. In other words, they are of about the same grade as English homozygotes. This means that a white Dutch gamete has about the same whitening effect on English as another gamete of English would have. The pattern of the English-Dutch heterozygote is indistinguishable either quantitatively or qualitatively from that of a homozygous English animal. The Dutch is not at all in evidence except as a whitening influence on the dominant English. Even in a single dose it completely counteracts the darkening influence introduced into the English race in the process of the cross with self and the subsequent 5 generations of selection.

Since it has been shown that both white Dutch and English segregate from self in monohybrid fashion, much interest attaches to the inquiry whether they segregate from each other in the gametes of the F_1 individuals. Two methods have been employed to test this matter, one being to mate F_1 individuals *inter se*, the other to backcross them with white Dutch. If English and Dutch are allelomorphs (invariably pass into different gametes) nothing but English or Dutch young should be produced by either mating. If, however, English and Dutch are not allelomorphs, then a certain number of gametes should be formed by .F_1 individuals which are *neither* English nor Dutch, but which are *self*, and a like number should be formed which carry *both* English and Dutch, the genetic properties of which would be to produce a very white (low-grade) English. These two new kinds of gametes would respectively be as numerous as the simple English and simple Dutch gametes, if English and Dutch are inherited independently (in different chromosomes, for example). If English and Dutch are linked in inheritance (are borne in homologous chromosomes but not at the same locus), these two classes of gametes would represent the "cross-overs." The existence of the gamete which bears both English and Dutch would be difficult to demonstrate, since one character is dominant, the other recessive; but the existence of gametes bearing *neither* English nor Dutch would be easy to detect, either in the straight F_2 generation or in the back-cross generation. Such gametes uniting with each other would produce *self* individuals, or uniting with a Dutch gamete would produce a Dutch of grade 3 or lower. Now, among 47 F_2 young neither of these classes of individuals has appeared. The 11 non-English F_2 young are all Dutch of grade 7 or higher. An even better test for the existence of gametes transmitting neither English nor Dutch is the back-cross with Dutch; for in this case any such gamete would produce one and the same kind of zygote, viz, Dutch of grade 3 or lower. No such zygote has appeared in a total of 88 Dutch and 105 English young obtained in back-cross matings. This indicates that English and Dutch are either allelomorphs or closely linked. The grade distribution of these back-cross young is shown in the table herewith.

It should be of interest to compare this distribution with that obtained in the original English-Dutch cross (page 26) as indicating whether the contrasted characters have been contaminated in the F_1 zygote. The English back-cross young are of lower grade (whiter) than the F_1 English. The respective means are 1.36 and 2.25.

This whitening of the English is the result of contamination from white Dutch in the F_1 zygote. No English individual produced in the back-cross is darker than the English produced in F_1. In both cases the darkest

Dutch young.		English young.	
Grade.	No.	Grade.	No.
6	2	1	52
7	1	2	25
8	1	3	2
9	4	?	26
10	6		
11	4		
12	13		
13	1		
14	15		
15	14		
16	22		
17	3		
?	2		
Total	88		105

English are of grade 3. This indicates absence of gametes transmitting *neither* English nor Dutch, for any such gametes uniting with an English gamete should produce English *darker* than grade 3. The F_1 Dutch (page 26) were self-white Dutch heterozygotes. (Compare table 27.) None was darker than grade 3. In this back-cross, any self gamete (neither Dutch nor English) which might arise by cross-over should produce young equally dark, grade 3 or lower. But the lowest-grade Dutch recorded are of grade 6;

hence there is no evidence that cross-over gametes are produced as often as once in 192 times. It, therefore appears that English and Dutch are either very closely coupled or are variants of the same locus. As regards the effect of the English cross on the grade of the Dutch character, this is indicated in the grade of the back-cross Dutch young, which range in grade from 6 to 17, average 13.88. Similarly produced back-cross Dutch obtained from F_1 (white × self) × white (table 17) range from 13 to 17 and are of mean grade 15.15. This indicates that the English cross has *darkened* white Dutch even more than self did in a similar cross (table 27). The superior darkening effects of English over the self used in table 27 may be attributed to the more highly selected character (for darkness) of the English used in the cross and to the less highly selected character (for whiteness) of the Dutch used in the same cross, but the difference is not great.

At any rate, from the grade distribution of the young produced in the back-cross of F_1 (English × white Dutch) × white Dutch, it is clear (1) that English and Dutch behave like allelomorphs, or closely linked factors, since in nearly 200 cases studied no cross-over is observed, *i.e.*, no gamete transmitting *neither* English nor Dutch; (2) that the segregated English and Dutch borne by the gametes of F_1 individuals are mutually modified, the English (previously selected for darkness) being made lighter, and the Dutch (previously selected for whiteness) being made darker. In these mutual modifications we are dealing probably for the most part with residual heredity, but it is possible that quantitative variation of the English and the Dutch genes is in part responsible, yet such an interpretation is not favored by the results obtained from the critical experiments with hooded rats, which point strongly to changed residual heredity as the correct explanation of changed phenotypes, when only a single Mendelizing character can be observed.

III. OBSERVATIONS ON THE OCCURRENCE OF LINKAGE IN RATS AND MICE.

In publication 241 of the Carnegie Institution evidence was presented showing that the red-eyed yellow and pink-eyed yellow variations of the common rat (*Mus norvegicus*) are due to genes which are linked with each other. Upon crossing with each other the two yellow variations, which visibly differ in eye-color only, young are obtained which differ from both parent races in coat-color as well as in eye-color. These young are black-coated or gray-coated and have black eyes. This result shows clearly that the two variations, which are both recessive in genetic behavior, are due to independent genes.

In the F_2 generation the two yellow varieties were recovered, each with its distinctive eye-color, and certain individuals, which visibly were pink-eyed yellows, were found from breeding tests to carry the genes for red-eyed as well as for pink-eyed yellow. These double recessives obviously had arisen by the process known as "crossing-over," in which genes, although introduced in a cross by different parents, yet later emerge together in the same gamete formed by an F_1 individual. It is supposed that genes which behave in this way lie in homologous chromosomes and that when crossing-over occurs a gene (A) leaves the chromosome in which it originally lay and crosses over into the homologous chromosome in which the other gene (B) lay. Thus both A and B come to lie in the same chromosome and at gametogenesis pass into the same gamete. From an examination of the proportion of the double recessive yellow individuals found among the F_2 yellows, it was concluded that cross-over gametes (those which carry genes for *both* yellow variations or for *neither*) represent about 18.5 per cent of all the gametes formed by F_1 individuals. If no linkage occurred, such gametes would form 50 per cent of the total.

To test more fully the strength of the linkage between these two genes and to find out whether this linkage has the same strength in both sexes, further experiments have been undertaken. A race of homozygous double recessives (genetically both pink-eyed and red-eyed) was built up from the F_2 cross-over individuals and with this race F_1 individuals were crossed. If we designate by r the gene for red-eyed yellow and by p the gene for pink-eyed yellow, an F_1 individual might be expected to form gametes of the four sorts PR, Pr, pR and pr. Of these 4 combinations, Pr and pR would correspond with those furnished by the parent races, red-eyed yellow and pink-eyed yellow respectively; but the other two, PR and pr, would be new and

29

could therefore arise only by crossing-over. A test mating of an F_1 individual with a double recessive would give zygotes as follows, it being understood that the double recessive individual produces only one type of gamete, viz. pr. Non-cross-over gametes, Pr and pR, would give zygotes Pprr and ppRr, visibly red-eyed yellows and pink-eyed yellows respectively. Cross-over gametes, PR and pr, would give zygotes PpRr and pprr, visibly dark-eyed (black or gray) and pink-eyed yellow respectively. The pink-eyed yellows could not be distinguished readily from yellows arising from non-cross-over gametes, but the dark-eyed young could be distinguished immediately at birth from all other classes. Since theoretically they would constitute half the total cross-overs, it is evident that the simplest way of estimating with accuracy the proportion of cross-over gametes is to double the number of dark-eyed young observed in new-born litters. This number divided by the total number of young would give the percentage of cross-over gametes.

Following this procedure, we have reared from matings of F_1 individuals with double recessives a total of 1,714 young, of which 174 were dark-eyed. Doubling the number 174, we have 348 as the probable number of cross-over gametes among the 1,714 F_1 gametes which entered into the production of these young. This is a percentage of 20.3, a little higher than the calculation 18.5 of publication 241, based on a study of a much smaller F_2 population. The difference between this figure, 20.3 and 50, the percentage expected where no linkage occurs, would be a measure of the strength of the *repulsion* shown between the genes for red-eyed yellow and for pink-eyed yellow respectively, when they enter a cross in different gametes—that is, each through a different parent, the condition realized in this cross.

But, on the chromosome theory, an attraction or "coupling" equal in strength to this repulsion should occur between the same two genes when they enter a cross together. Entering together, they should tend to *hold together*, because they would lie in the same member of a pair of chromosomes and so could pass out separately only in consequence of a cross-over. This point, repeatedly verified in the case of other organisms, was tested for rats by producing F_1 individuals through a cross of double recessive yellow (pprr) with a pure non-yellow individual (PPRR). In reality F_1 zygotes of this same sort were being produced in considerable numbers in the matings already described to test the strength of repulsion. Such were the 174 dark-eyed young already mentioned. Each resulted from the union of a pr gamete with a PR gamete, the relationship which would give the expected coupling. Accordingly many of these dark-eyed young were used instead of F_1 parents in the experiments to test the strength of "coupling" between p and r. In these experiments, as in those to test the strength of repulsion, F_1 individuals were mated with double recessives. In both cases the F_1 individual was of the formula PpRr, the double recessive was of the formula pprr.

The only difference in the two cases was that in one case the F_1 arose from a union of Pr with pR, and in the other from a union of pr with PR. But the importance of this circumstance is seen in the different results obtained in the two cases. In one case (where p and r enter the cross separately) 10 per cent of the young were dark, in the other case (where p and r enter the cross together) more than 40 per cent of the young were dark. The exact figures for the repulsion series have already been given, 174 dark young in a total of 1,714, or 10.1 per cent dark young. For the coupling series the figures are 1,255 dark young in a total of 3,032, or 41.3 per cent dark young. To compare the *strength of repulsion* with the *strength of coupling* we may estimate the percentage of cross-over gametes produced in each case. Either sort of F_1 individual would produce 4 kinds of gametes, PR, Pr, pR, and pr. But in the repulsion series PR and pr would arise from crossing-over, whereas in the coupling series Pr and pR would arise from crossing-over. In either case dark-eyed individuals would arise only from the same type of F_1 gamete, viz, PR, but in the repulsion series this would be a cross-over gamete, whereas in the coupling series it would be a non-cross-over gamete. While in the repulsion series the number of dark-eyed young would measure half the total number of cross-overs, in the coupling series it would measure half the total number of non-cross-overs. Applying these criteria, we have found in the repulsion series, as already stated, that the number of dark-eyed young being 174, the probable number of cross-over gametes is twice this, or 348, in a total of 1,714, which is 20.3 per cent.

Turning now to the coupling series, we find that the total number of dark-eyed young is 1,255. Doubling this we have 2,510 as the probable number of non-cross-over gametes. Deducting this number from 3,032, the total number of young, we have 522 as the probable number of cross-over gametes, which is 17.2 per cent. This we may compare with the 20.3 estimated for the repulsion series, and the earlier estimate of 18.5 based on the census of an F_2 population (publication 241). These differences are not large enough to lead us to think that there is any consistent difference between the strength of repulsion and the strength of coupling between the same two genes. The chromosome theory would not lead us to expect the existence of any such difference. This case therefore fully accords with that theory. If we combine the results obtained from both the repulsion and the coupling series we have as the average *linkage strength* (either repulsion or coupling, as the case may be) 18.3 per cent. This is based on a total of 4,746 young produced by the back-cross of F_1 with the double recessive. The figures are large enough to have significance and agree remarkably well (almost too well) with the estimate based on the F_2 population, viz, 18.5 per cent.   It is safe to conclude that the linkage strength of red-eyed yellow with pink-eyed yellow is close to 18 per cent.

We may pass now to the question whether the linkage strength is the same in spermatogenesis as in oögenesis, whether it is the same among the gametes formed by F_1 males as in those formed by F_1 females. *A priori* we might well expect it to be different in the two cases, since in *Drosophila* crossing-over has been found to occur only in females, whereas in the silkworm it has been found to occur only in males. In publication 241 the fact was demonstrated that crossing-over does occur in both sexes of the rat, but we were at that time unable to state what its relative frequency was in the two sexes. Our back-cross series of matings give data for such a determination. (See table 48.) It will be observed that the estimated percentage of cross-over gametes is somewhat higher for females than for males in both the repulsion and the coupling series and that the difference is greatest where the numbers are largest, viz, in the coupling series. This would suggest that crossing over occurs more readily in oögenesis than in spermatogenesis, but I doubt very much whether such is the case when all other conditions are the same. Summaries made for the concluding period of our experimental work, when conditions had been more carefully controlled and the procedure of taking the records had been best standardized, show no appreciable differences in the case of the two sexes. For this period, in the coupling series, F_1 females gave 187 dark and 277 yellow young, or 19.3 per cent cross-over gametes. Simultaneously, F_1 males of similar parentage gave 133 dark and 197 yellow young, or 19.4 per cent cross-over gametes. In the repulsion series only F_1 males were at this time being used to any great extent. They produced 29 dark young and 269 yellow young, which by the method of calculation already explained indicates 19.4 per cent cross-over gametes, a remarkably close agreement with the results given by both sexes in the coupling series at this same period.

Whether external conditions have any influence on the percentage of cross-overs we are unable to state, but this seems doubtful in the case of a warm-blooded animal such as the rat. That individual or age differences may occur among F_1 animals affecting the percentage of cross-overs is a possibility we have considered carefully, but with only negative conclusions. The indicated percentage of cross-overs varies in the case of particular F_1 males from 0 to 44 per cent, but this variation appears to be the result of random sampling rather than of consistent differences in genetic behavior. Several males showing extremely high or extremely low percentage of cross-overs were transferred to new breeding-pens and mated with other double recessive females. Their indicated percentages of cross-over gametes before and after the transfer showed no consistency with each other, and so we are forced to conclude that individual differences as regards the production of many or few cross-overs have not been shown to exist.

In publication 241 evidence was presented indicating that albinism in rats is probably due to a gene which is linked with the genes for red-eyed yellow and pink-eyed yellow. This idea is now fully established and we are able to give provisional estimates of the linkage strengths involved, although the investigation of this matter is still incomplete.

When red-eyed and pink-eyed rats are crossed with each other, or either sort is crossed with an albino, the F_1 young produced are dark-eyed and dark-coated (either black or gray, according as the agouti factor is absent or present). But the F_1 young are not quite as dark in color as wild rats. This shows that all three variations are recessive and complementary, but that the allelomorph of each is a little less effective in producing pigment when in heterozygous form than when in homozygous form (as in wild rats, or in Irish or in hooded rats). F_1 individuals from the cross of albino with pink-eyed yellow, when bred with each other, produce an F_2 generation of three apparent types, viz, (1) dark, in eye and coat color; (2) pink-eyed yellow; (3) albino. If no linkage occurred we should expect these three classes to occur in the ratio $9:3:4$; but, as was pointed out in publication 241, linkage would tend to equalize the numbers of pink-eyed and albino young, and such a tendency has been recorded. Further, if no linkage occurs, but if pink-eyed yellow and albinism segregate quite independently of each other, then half the albino gametes formed by F_1 individuals should transmit pink-eyed yellow and half should not transmit it; conversely, half the gametes which transmit pink-eyed yellow should also transmit albinism and half should not. If less than half the gametes which transmit one character transmit the other, the two show repulsion.

To test the matter, 45 F_2 albinos have been mated with homozygous pink-eyed individuals. Of the 45 so mated, 17 have produced both pink-eyed and dark-eyed young, one has produced only pink-eyed young, and 27 have produced only dark-eyed young. The 17 are evidently heterozygous for pink-eye as well as homozygous for albinism, their formula being ccpP. The one which produced only pink-eyed young is probably of the formula ccpp. The 27 which produced only dark-eyed young are of the formula ccPP.

We may now consider what was the nature of the gametes which produced these 45 individuals. A gamete which furnished both albinism and pink may be called a cross-over gamete; one which furnished albinism only must be regarded as a non-cross-over gamete. The 27 albinos which did not transmit pink-eye evidently arose each from the union of two non-cross-over gametes. This accounts for 2×27 or 54 non-cross-over gametes. The 17 individuals which were heterozygous for pink evidently received each a single non-cross-over gamete. This makes a total of 71 such gametes. Cross-over gametes were represented singly in each of the 17 individuals which were heterozygous for pink and doubly in the one which was homozygous for pink. This makes a total of 19 cross-over as against 71 non-cross-over gametes, which is 21.1 per cent cross-overs. This is an indicated linkage strength a little less close than that between pink-eyed and red-eyed yellow, in which case the percentage of cross-over gametes was estimated at 18.3. For with *no-linkage* giving 50 per cent cross-overs, it is evident that the linkage strength increases as the percentage of cross-overs decreases until (when cross-overs cease) linkage becomes complete. If we measure the strength of linkage by the difference between the observed percentage of cross-overs and 50 per cent (the percentage of cross-overs when no linkage occurs), then linkage between red-eyed yellow and pink-eyed yellow is 31.7 and that between pink-eyed yellow and albinism is 28.9, as provisionally determined.

The linkage between red-eyed yellow and albinism is much stronger than the linkage in either of the cases just discussed. Tests have been made for the presence of the red-eyed yellow gene in 160 F_2 albinos and for the presence of albinism in 57 F_2 red-eyed yellows derived from the cross of albino with red-eyed yellow. Only a single cross-over has been detected, and even that is not beyond question. One of the F_2 albinos, a male, when mated with a pure red-eyed yellow female, sired a litter of young, all of which were dark-eyed except one. This one proved to be a yellow but died, as did the father, before additional breeding tests could be applied. If this yellow individual was really sired by the albino male (and not accidentally introduced from some other cage, a remote possibility), then that male evidently carried yellow as well as albinism and in his genesis a cross-over gamete must have functioned. Each of the other F_2 albinos and the F_2 yellows tested manifestly arose from the union of gametes neither of which transmitted both yellow and albinism, since as mated they produced only dark-eyed young (4 or more each). On these assumptions the experiments thus far show that only one gamete out of 434 formed by F_1 parents can have been a cross-over gamete, which apparently gives less than one per cent of cross-overs.

The experiments are being continued with the hope of finding additional cross-overs and of thus securing a double recessive race, which will make possible a more accurate determination of the linkage strength. The information already presented shows that on the chromosome theory the genes for albinism and for red-eyed yellow are extremely close to each other in the same chromosome and that the gene for pink-eyed yellow, while lying in this same chromosome, is at some distance from the genes for albinism and red-eyed yellow.

In mice it has been shown by Haldane *et al.* that the genes for albinism and for pink-eye are probably linked with each other. This fact, interesting in itself, is made doubly so by the consideration that characters apparently identical in nature with these two are also linked with each other in rats. Since mouse and rat are species grouped by systematists in a single genus, it should be of interest to compare their genetic constitution as fully as possible. With this idea in mind we had already undertaken to study the linkage relations of albinism and pink-eye in mice before the appearance of the paper by Haldane *et al.*[1] This investigation was undertaken by Mr. L. C. Dunn while acting as my assistant. Upon his entering military service, I took over the experiments. It is a pleasure to acknowledge Mr. Dunn's important part in the work.

We began, as in the rat experiments, by crossing pink-eyed with albino individuals. Dark-eyed F_1 young were produced exactly as in rats. These bred with each other produced an F_2 generation of dark-eyed young, pink-eyed young, and albinos, in a 9 : 3 : 4 ratio manifestly modified by linkage. Pink-eyed F_2 individuals were tested for the presence of albinism and albino F_2 individuals were tested for the presence of the pink-eye gene as a first step toward the production of a race of double recessives needed to ascertain the proportion of cross-over gametes formed by F_1 individuals. The simplest way of making the tests was found to be the mating of F_2 albinos with F_2 pink-eyed individuals. This afforded simultaneously a test of both parents. For if the pink-eyed parent carried albinism, 50 per cent of the young would be albinos, otherwise none would be albinos. But if the albino carried the pink-eye gene, 50 per cent of the young produced would be pink-eyed. If both these contingencies were realized in the mating, 25 per cent of the young would be pink-eyed and 25 per cent albinos. All other young, as in a cross of pure pink-eyed with pure albinos, would be dark-eyed.

If no linkage occurred between pink-eye and albinism, it would be expected that half the F_2 pink-eyed individuals would carry albinism, and also that half the albinos would carry pink-eye. Any smaller proportions than these of pink-eyed carrying albinism or of albinos carrying pink-eye, among the F_2 individuals, would indicate linkage.

Linkage is very clearly shown by the tests made. Among 63 F_2 pink-eyed which were tested, 18 produced each one or more albino young in litters otherwise dark-eyed, while 45 produced no albinos but only dark-eyed young. In the genesis of the 18 parents mentioned, it is evident that 18 cross-over gametes had united with 18

---

[1] In fact, I have not yet had access to the paper by Haldane *et al.* but know it only as cited by others. Our copy of the journal in which it appeared is probably at the bottom of the ocean and we have been unable as yet to replace it.

non-cross-overs. But in the production of the 45, only non-cross-over gametes had functioned. The total gametes involved accordingly are 18 cross-overs and $18 + (2 \times 45) = 108$ non-cross-over gametes; total 126. As 18 is 14.28 per cent of 126, the indicated percentage of cross-overs is 14.28 per cent.

Among 75 F_2 albinos which were tested, 20 produced pink-eyed young (as well as dark-eyed ones), while the remaining 55 produced only dark-eyed young. Reasoning as before, there were evidently involved in this case 20 cross-over gametes and $20 + (2 \times 55) = 130$ non-cross-overs, total 150. But 20 is $13\frac{1}{3}$ per cent of 150; hence the indicated percentage of cross-overs is $13\frac{1}{3}$.

Combining the tests of F_2 pink-eyed and of F_2 albinos, we have in tests involving 276 F_1 gametes an indicated percentage of 13.76 cross-overs.

The pink-eyed parents, which in the course of these tests had been found to carry albinism, were now mated with each other, and the albino young which they produced when so mated were used in building up a race of double recessives, for all albinos so produced must of necessity be homozygous for pink-eye. These double recessives were next mated with F_1 dark-eyed animals obtained by the original cross of pink-eyed with albino, or with dark-eyed individuals of similar genetic constitution which had resulted from the test matings. The interpretation of the results obtained from these back-cross matings is the same as that given by similar matings in the case of rats. The F_1 parent would form gametes of the four sorts CP, cP, Cp, and cp, of which cP and Cp would represent the original combinations found in pure albinos and pure pink-eyed respectively, and so would be *non-cross-overs*, but CP and cp would arise only by crossing-over. Of the four types of gamete, CP alone would produce a dark-eyed zygote, if mated with a double recessive, cp. But this is one of the two cross-over types. Hence the number of dark-eyed young produced in mating F_1 animals with double recessives should indicate half the total percentage of cross-overs. By matings of the sort just described, 3,142 young have been produced, of which 222 were dark-eyed. Doubling this number, we have 444 as the probable number of cross-over gametes in 3,142 gametes produced by the F_1 parents, an indicated percentage of 14.13. This agrees very well indeed with the 13.76 per cent indicated by the test-matings of F_2 pink-eyed and albinos. It seems safe to assume, therefore, that the average cross-over percentage is close to 14 per cent. For the corresponding characters in rats, the indicated percentage of cross-overs is considerably higher, viz, 21.1, but it should be borne in mind that the estimate is based on a much smaller series of observations in the case of rats and that further observations may alter it materially.

TABLES.

TABLE 1.—*Classification of generation 17, plus-selection series.*

Grade of parents	Grade of offspring.									Totals.	Means.
	3¾	4	4¼	4½	4¾	5	5¼	5½	5¾		
4⅛	2	2	1	..	..	..	..	..	..	5	3.95
4¼	1	2	3	3	..	..	..	..	.	9	4.22
4⅜	..	1	5	4	1	..	1	..	..	12	4.44
4½	..	4	11	7	4	4	1	..	..	31	4.47
4⅝	1	7	7	11	6	2	4	..	..	38	4.49
4¾	5	12	26	17	13	8	6	1	..	88	4.46
4⅞	1	8	10	13	5	1	3	2	..	43	4 48
5	9	5	10	12	10	10	4	5	..	65	4.56
5⅛	3	2	13	6	2	5	3	3	2	39	4.61
5¼	1	3	8	1	2	1	..	..	..	16	4.30
5⅜	..	..	2	1	1	1	..	..	..	5	4.55
4.81	23	46	96	75	44	32	22	11	2	351	4.48

TABLE 2.—*Classification of generation 18, plus-selection series.*

Grade of parents.	Grade of offspring.									Totals.	Means.
	3½	3¾	4	4¼	4½	4¾	5	5¼	5½		
4¼	3	4	5	3	1	..	2	1	..	19	4.10
4⅜	..	..	9	9	9	3	2	..	..	32	3.34
4½	..	..	3	8	10	7	5	1	1	35	4.57
4⅝	..	5	10	22	27	12	4	5	1	86	4.45
4¾	..	2	3	13	7	6	8	6	3	48	4.64
4⅞	..	5	8	14	5	2	1	..	..	35	4.21
5	..	7	11	25	16	12	9	3	1	84	4.43
5⅛	..	2	4	11	14	2	7	2	1	43	4.50
5¼	..	1	2	7	5	3	4	5	..	27	4.61
5½	..	..	..	2	3	2	..	2	2	11	4.82
4.80	3	26	55	114	97	49	42	25	9	420	4.46

TABLE 3.—*Classification of generation 19, plus-selection series.*

Grade of parents.	Grade of offspring.									Totals	Means.
	3½	3¾	4	4¼	4½	4¾	5	5¼	5½		
4¼	..	..	2	4	1	..	..	..	..	0	4.21
4⅜	3	8	15	25	23	9	7	4	1	95	4.34
4½	..	..	..	4	4	3	2	..	..	13	4.56
4⅝	..	..	3	4	5	2	1	2	..	17	4.50
4¾	..	4	6	15	20	8	8	3	1	65	4.49
4⅞	..	1	..	6	12	10	8	8	2	47	4.76
5	..	3	1	4	7	3	3	1	..	22	4.47
5¼	..	..	..	3	6	3	2	..	..	14	4.53
4.66	3	16	27	65	78	38	31	18	4	280	4.49

TABLE 4.—*Classification of generation 20, plus-selection series.*

Grade of parents.	Grade of offspring.									Totals.	Means.
	$3\frac{3}{4}$	4	$4\frac{1}{4}$	$4\frac{1}{2}$	$4\frac{3}{4}$	5	$5\frac{1}{4}$	$5\frac{1}{2}$	$5\frac{3}{4}$		
$4\frac{1}{4}$	1	2	3	3	..	..	..	..	..	9	4.22
$4\frac{3}{8}$	..	1	5	5	2	1	1	..	..	15	4.50
$4\frac{1}{2}$	..	2	9	3	3	2	..	..	..	19	4.42
$4\frac{5}{8}$	..	..	1	4	1	1	..	..	..	7	4.57
$4\frac{3}{4}$	..	1	1	3	3	3	1	2	..	14	4.77
$4\frac{7}{8}$	..	..	1	3	1	2	3	1	2	13	5.02
5	..	..	..	7	..	2	..	..	..	9	4.61
$5\frac{1}{4}$	..	..	1	..	..	..	2	..	..	3	4.89
$5\frac{1}{2}$	..	..	..	..	3	..	..	..	..	3	4.75
4.66	1	6	21	28	13	11	7	3	2	92	4.61

TABLE 5.—*Classification of generation 18, minus-selection series.*

Grade of parents.	Grade of offspring (minus).								Totals.	Means.
	$2\frac{1}{4}$	$2\frac{1}{2}$	$2\frac{3}{4}$	3	$3\frac{1}{4}$	$3\frac{1}{2}$	$3\frac{3}{4}$	4		
$-2\frac{3}{4}$	3	26	29	2	1	..	..	..	61	−2.63
$-2\frac{7}{8}$	3	5	28	9	4	..	..	..	49	−2.78
-3	2	18	37	9	2	..	..	..	68	−2.72
$-3\frac{1}{8}$	..	5	19	13	5	4	..	1	47	−2.94
$-3\frac{1}{4}$	..	1	15	8	8	1	1	..	34	−2.97
$-3\frac{3}{8}$	..	..	12	8	6	1	..	1	28	−3.00
$-3\frac{1}{2}$	..	..	3	1	2	3	..	..	9	−3.14
$-3\frac{5}{8}$	..	4	9	4	5	6	1	..	29	−3.03
$-3\frac{7}{8}$	..	..	..	1	1	1	2	..	5	−3.45
−3.09	8	59	152	55	34	16	4	2	330	−2.84

TABLE 6.—*Classification of generation 19, minus-selection series.*

Grade of parents.	Grade of offspring (minus).							Totals.	Means.
	$2\frac{1}{4}$	$2\frac{1}{2}$	$2\frac{3}{4}$	3	$3\frac{1}{4}$	$3\frac{1}{2}$	4		
$-2\frac{3}{8}$	..	1	2	2	..	..	..	5	−2.80
$-2\frac{7}{8}$	..	5	11	3	1	1	..	21	−2.79
-3	3	4	16	7	3	2	1	36	−2.85
$-3\frac{1}{8}$	..	3	11	15	7	1	..	37	−2.95
$-3\frac{1}{4}$	..	..	7	2	2	..	..	11	−2.89
$-3\frac{3}{8}$	..	..	2	7	2	2	..	13	−3.08
$-3\frac{1}{2}$	..	2	1	..	..	..	..	3	......
$-3\frac{3}{4}$	..	..	..	2	1	..	..	3	......
$-3\frac{7}{8}$	..	..	..	..	1	..	..	1	......
−3.10	3	15	50	38	17	6	1	130	−2.89

TABLE 7.—*Classification of generation 20, minus-selection series.*

Grade of parents.	Grade of offspring (minus).							Totals.	Means.
	2	$2\frac{1}{4}$	$2\frac{1}{2}$	$2\frac{3}{4}$	3	$3\frac{1}{4}$	$3\frac{1}{2}$		
$-2\frac{1}{2}$	..	1	8	9	3	..	..	21	−2.67
$-2\frac{3}{4}$	..	1	..	2	1	1	..	5	−2.80
$-2\frac{7}{8}$	1	..	6	21	6	6	..	40	−2.81
-3	..	..	1	2	4	..	..	7	−2.86
$-3\frac{1}{4}$	..	..	..	5	..	..	1	6	−2.87
−2.81	1	2	15	39	14	7	1	79	−2.78

TABLE 8.—*Classification of generation 21, minus-selection series.*

Grade of parents.	Grade of offspring (minus).							Totals.	Means.
	2	$2\frac{1}{4}$	$2\frac{1}{2}$	$2\frac{3}{4}$	3	$3\frac{1}{4}$	$3\frac{1}{2}$		
$-2\frac{5}{8}$	1	2	7	2	..	..	2	14	−2.61
$-2\frac{3}{4}$	..	1	4	1	..	1	..	7	−2.61
$-2\frac{7}{8}$	..	..	2	1	..	..	..	3	......
-3	..	..	..	1	..	..	..	1	......
$-3\frac{1}{4}$	..	..	1	2	3	1	3	10	−3.11
−2.58	1	3	14	7	3	2	5	35	−2.74

TABLE 9.—*Classification of extracted hooded third F_2 young produced by a cross of plus-selected with wild rats.*

2d F_2 grand-parent.	2	$2\frac{1}{4}$	$2\frac{1}{2}$	$2\frac{3}{4}$	3	$3\frac{1}{4}$	$3\frac{1}{2}$	Totals.	Means.	σ
♀ 208, $+2\frac{3}{4}$..........	..	..	..	..	..	..	1	1	....	..
♀ 9922, $+3$..........	2	1	2	..	1	1	2	9	....	..
♂ 63, $+3\frac{1}{2}$..........	..	..	..	2	..	1	6	9	....	..
Weighted mean, 3.22.	2	1	2	2	1	2	9	19	3.04	.64

NOTE. Standard deviation of *first* F_2 was 0.73, that of *second* F_2 was 0.50 (tables 141 and 145, Publication No. 241).

TABLE 10.—*Classification of extracted hooded first F_2 young from a cross of minus-selected with wild rats.*

Hooded grand-parent.	$2\frac{1}{4}$	2	$1\frac{3}{4}$	$1\frac{1}{2}$	$1\frac{1}{4}$	1	$\frac{3}{4}$	$\frac{1}{2}$	$-\frac{1}{4}$	0	$+\frac{1}{4}$	$\frac{1}{2}$	$\frac{3}{4}$	1	$1\frac{1}{4}$	$1\frac{1}{2}$	$1\frac{3}{4}$	2	$2\frac{1}{4}$	$2\frac{1}{2}$	$2\frac{3}{4}$	3	Total.	Means.	σ
♀ 20,331, $-2\frac{3}{4}$, gen. $15\frac{1}{4}$	..	1	..	1	4	1	..	1	8	..	2	.	4	..	1	1	2	.	.	2	1	.	29	+.46	...
♀ 20,482, $-2\frac{3}{4}$, gen. $15\frac{3}{4}$	..	3	.	1	..	4	.	2	1	1	..	1	.	1	1	.	.	.	.	.	.	1	16	−.41	...
♀ 20,359, $-2\frac{3}{4}$, gen. 16	..	2	2	..	1	..	.	.	.	.	.	.	.	.	.	.	.	.	.	.	.	.	5	......	...
♀ 20,327, $-2\frac{3}{4}$, gen. $15\frac{1}{2}$	2	2	7	6	..	13	.	7	..	.	7	..	..	1	4	..	2	1	1	..	.	.	53	−.65	...
♀ 20,480, $-2\frac{3}{4}$, gen. $15\frac{3}{4}$	..	6	..	.	4	.	1	..	3	.	.	.	.	.	3	1	.	.	.	.	.	.	18	−.49	...
Total..............	2	7	16	7	1	26	1	10	2	19	..	3	1	8	1	7	3	3	..	2	1	1	121	−.38	1.25

TABLE 11.—*Classification of extracted hooded second F_2 young from the cross of minus-selected with wild rats.*

From original hooded.	1st F_2 grand-parent.	1¾	1½	1¼	1	¾	½	−¼	0	+¼	½	¾	1	1¼	1½	1¾	2	2¼	2½	2¾	Totals.	Means.	σ
♀ 20,480.	♀ 1698, −1¾	..	..	..	..	..	..	.	2	...	2	1	1	4	1	1	1	..	1	..	14	1.11	...
♀ 20,327.	♀ 1715, −1¾	..	..	..	..	1	.	1	...	4	...	2	1	2	..	3	1	.	1	..	17	.80	...
♀ 20,327..	♀ 1563, −1½	1	..	..	..	1	...	3	.	1	1	...	..	2	..	..	..	..	..	..	9	.67	...
♀ 20,480.	♀ 944, +1½	..	..	..	..	..	..	.	..	1	.	1	..	2	2	2	..	..	1	..	9	1.64	...
	Weighted mean, −1.11	1	..	..	..	1	.	1	...	7	2	8	2	5	5	6	4	5	1	1	49	1.01	.92

TABLE 12.—*Classification of extracted hooded third F_2 young from a cross of minus-selected with wild rats.*

From 1st F_2.	From 2d F_2.	1	1¼	1½	1¾	2	2¼	2½	2¾	3	3¼	3½	Totals.	Means.	σ
♀ 944, −1½	♀ 1924, +2	3	..	1	3	2	2	9	8	8	4	1	41	2.52	..
♀ 944, +1½	♀ 1925, +½	1	..	2	1	2	2	3	3	6	3	..	23	2.51	..
♀ 944, +1½	♂ 1926, +1½		1	..	1	..	..	1	5	2	3	..	13	2.68	..
♀ 1563, −1½	♀ 2008, +½	1	..	..	2	1	..	..	1	..	2	..	7	2.25	..
♀ 1698, −1¾	♂ 2048, +1¾	1	..	2	1	1	..	..	..	1	..	..	6	1.79	..
♀ 1563, −1½	♂ 2068, +3	..	..	..	..	1	1	2	4	2	4		14	3.05	..
	Weighted mean, +1.62	6	1	5	8	6	5	14	19	21	14	5	104	2.55	.66

TABLE 13.—*Grade distribution of F_1 young sired by the standard-bred Dutch ♂ 3037 (grade 7) mated with does transmitting the self (unspotted) condition.*

Mothers designated (E) were English marked and transmitted the self condition in only half of their gametes. The English young of such mothers are omitted from this table. Mothers designated (H) were Himalayan albinos.

Parents.	Grades of young.				Totals	Mean grade.
	0	1	2	3		
♂3037(7²) × ♀ 2651(E)	.	1	1	1	1	} 2.17
" × ♀ 2688(E)	..	1	..	2	3	
" × ♀ 2687(H)	..	4	1	..	5	
" × ♀ 2830(H)	3	5	1	..	9	} .95
" × ♀ 2835(H)	..	4	..	..	4	
Totals	3	15	3	3	24	1.25

TABLE 14.—*Grade distribution of the back-cross young produced by F_1 does (table 13) mated with the standard-bred Dutch buck ♂ 3036 (grade 9).*

Parents.	Grades of young.							Totals.	Mean.
	1	2	3	4	5	6	7		
♂3036(9²) × ♀ 5001(1)	2	4	..	2	1	..	..	9	....
" × ♀ 5032(1)	..	..	..	1	.	1	3	5	....
" × ♀ 5003(3)	..	..	..	..	..	4	2	6	....
Totals	2	4	..	3	1	5	5	20	4.60

² The number in parentheses indicates the grade (pl. 1) of the animal.

TABLE 15.—*Grade distribution of F₂ young from the cross indicated in table 13.*

Parents.	0	1	2	3	4	5	Totals.	Mean grade.
♂5002(3) × ♀5001(1).....	..	3	2	..	1	2	8	}2.45
" × ♀5032(1)......	..	3	3	1	1	..	8	
" × ♀5003(3).....	..	1	5	5	2	..	13	
♂5029(1) × ♀5001(1)......	7	14	3	1	..	..	25	}1.34
" × ♀5032(1)......	1	4	..	3	..	..	8	
" × ♀5003(3)......	..	2	3	2	..	1	8	
Totals...............	8	27	16	12	4	3	70	1.80

TABLE 16.—*Grade distribution of young produced by F₂ does (table 15) or back-cross does (table 14) mated with the standard-bred buck, 3036 (table 14).*

Parents.	1	2	3	4	5	6	7	8	9	10	11	12	13	14	15	16	17	Totals.	Mean lower group.	Mean higher group.
♂3036(9) × ♀5150(5), F₂ ..	.	.	.	.	2	3	2	.	1									8	}4.94	....
" × ♀5153(5), F₂	.	1	4	11	1	5	.	.										22		
" × ♀5170(6), BC	.	3	.	5	3	4	.	.	2	1								18	}5.27	14.55
" × ♀5166(4), BC	1	.	1	.	.	.	1	.	.				1	2				6		
" × ♀5158(5), BC	.	.	.	3	4	.	.	1	.	1			1	2				12		
" × ♀5159(6), BC	.	.	.	2	1	2	.	.						2				7		
" × ♀5169(6), BC	.	.	.	3	3	.	1	3	2	.	.	.	1	.	.	.	1	14		
Totals..	1	4	5	24	14	16	6	2	4	1			1	2	6		1	87	5.06	14.55

TABLE 17.—*Grade distribution of young produced by second back-cross does recorded in table 16 and the same standard-bred buck, 3036 (tables 14 and 16).*

Parents.	1	2	3	4	5	6	7	8	9	10	11	12	13	14	15	16	17	Totals.	Means, lower group.	Means, upper group.
♂3036(9) × ♀5536(7), 2BC.	.	.	.	.	.	.	2	2	.	.	.	.	.	.	.	.	..	4	7.50	....
" × ♀5590(9), 2BC.	.	.	.	1	2	1	5	2	3	.	1	.	1	3	2	.	.	21	8.00	15.17
Totals................				1	2	3	7	2	3	.	1	.	1	3	2	.	.	25	7.89	15.17

TABLE 18.—*Grade distribution of young produced by F₂, back-cross and second back-cross does (tables 14–16) mated with a back-cross buck, ♂5167, grade 7 (table 14), son of ♂5003(3).*

Parents.	2	3	4	5	6	7	8	9	10	11	12	13	14	15	16	17	Totals.	Means, lower group.	Means, upper group.
♂5167(7) × ♀5150(5), F₂....	.	.	.	4	1	.	.	.	.	.	.	.	.	.	.	.	5	}5.11	
" × ♀5153(3), F₂.....	.	1	1	.	1	1	.	.	.	.	.	.	.	.	.	.	4		
" × ♀5166(4), BC...	1	.	2	1	.	.	.	.	.	.	.	.	3	2	.	.	9	}4.00	14.40
" × ♀5158(5), BC...	.	1	2	2	.	.	.	1	.	.	.	.	.	.	.	.	6		
" × ♀5601(6), 2BC...	.	.	.	3	.	.	.	.	.	.	.	.	2	.	.	1	6		
" × ♀5645(6), 2BC...	.	2	.	3	3	.	.	.	.	.	.	.	.	.	.	.	8		
" × ♀5920(6), 2BC...	1	1	1	2	.	1	.	.	.	.	.	.	.	.	.	.	6	}4.85	15 67
" × ♀5933(3), 2BC...	.	.	1	4	.	.	.	.	.	.	.	.	.	.	.	.	5		
" × ♀5936(5), 2BC...	.	.	.	2	2	.	.	.	.	.	.	.	.	.	.	.	4		
Totals.................	2	5	7	21	7	2	.	1	.	.	.	.	3	4	.	1	53	4.73	14.87

TABLE 19.—*Variation of the uncrossed "white" Dutch race.*

Parents.	Grades of young.			Totals.	Means.
	15	16	17		
♂6175(17) × ♀5945(15)........	1	..	..	1	
" × ♀7934(15)........	2	4	5	11	} 16.17
" × ♀6703(16)........	3	15	7	25	
" × ♀7003(16)........	2	4.	1	7	} 16.12
" × ♀7185(16)........	..	..	1	1	
" × ♀7313(17)........	1	..	6	7	16.71
♂9218(17) × ♀9222(15)........	..	..	3	3	17.00
" × ♀9217(16)........	..	1	3	4	16.75
Totals.....................	9	24	26	59	16.25

TABLE 20.—*Variation of the uncrossed "dark" Dutch race.*

Parents.	Grades of young.							Totals.	Means.
	1	2	3	4	5	6	7		
♂6701(5) × ♀7641(2)......	2	5	6	1	3	1	..	18	} 3.05
" × ♀7642(2)......	..	2	..	..	1	..	..	3	
" × ♀8034(3)......	..	..	4	2	2	4	..	12	
" × ♀7644(3)......	..	1	..	3	..	..	..	4	} 3.35
" × ♀7684(3)......	3	7	10	5	..	1	..	26	
" × ♀6989(4)......	3	15	10	7	4	..	..	39	
" × ♀7685(4)......	..	8	6	9	3	1	..	27	} 3.35
" × ♀8290(4)......	..	2	8	8	8	1	1	28	
" × ♀5153(5)......	..	4	3	3	..	..	..	10	} 3.20
" × ♀6707(5) ...	..	..	3	1	..	1	..	5	
Totals.................	8	44	50	39	21	9	1	172	3.30

TABLE 21.—*Variation of the uncrossed "tan" Dutch race.*

Parents.	Grades of young.				Totals.	Means.
	2	3	4	5		
♂5757(3) × ♀7393(3)........	..	2	2	..	4	3.50
♂7142(4) × ♀9275(2)........	2	8	1	..	11	2.91
" × ♀9608(3).	2	3	1	..	6	
" × ♀6424(3).... .	..	2	..	1	3	
" × ♀8881(3)..	..	3	4	..	7	} 3.40
" × ♀9044(3)........	..	..	1	1	2	
♂6240(4) × ♀8881(3)........	1	3	3	..	7	
Totals.................	5	21	12	2	40	3.27

TABLE 22.—*Grade distribution of the F₁ young produced by the cross of "white" with "dark" Dutch.*

Parents.	Grades of young.							Totals.	Means.
	5	6	7	8	9	10	11		
♂6175(17) × ♀6666(5)......	1	1	4	2	.	1	1	10	
" × ♀5153(5)......	1	1	3	1	.	..	..	6	} 7.28
" × ♀6705(5)......	..	1	2	1	1	..	..	5	
" × ♀9170(6)......	..	3	..	1	1	..	..	5	} 7.28
" × ♀6038(8).. ...	..	.	.	2	.	..	..	2	
Totals..................	2	6	9	7	2	1	1	28	7.28

TABLE 23.—*Grade distribution of the F₁ young produced by the cross of "white" with heterozygous "dark" Dutch.*

Parents.	Grades of young.														Totals.	Means, lower group.	Means, higher group.
	4	5	6	7	8	9	10	11	12	13	14	15	16	17			
♂6175(17) × ♀5158(5) ...	1	1	1	2	..	..	..	..	..	..	1	3	..	1	10	} 6.33	15.55
" × ♀5940(5)....	..	2	1	1	3	..	..	..	..	..	..	..	4	..	11		
" × ♀5920(6)....	..	..	..	..	1	1	..	..	..	..	..	3	1	..	6		
" × ♀5601(6)....	..	..	..	1	..	1	..	..	..	..	..	..	2	1	5	} 6.82	15.67
" × ♀5939(6)....	..	..	..	..	1	1	..	..	..	..	..	3	..	..	5		
" × ♀6570(6)....	2	3	1	2	2	..	..	..	..	..	..	..	1	1	12		
" × ♀6643(7)....	..	..	1	2	4	1	..	..	..	..	..	1	2	3	14	} 7.57	16.25
" × ♀6891(7)....	..	..	1	2	2	1	..	..	..	..	..	1	6	3	16		
" × ♀6031(8)....	1	..	2	..	1	2	..	..	..	..	2	2	4	1	15	} 7.22	15.50
" × ♀7118(8)....	..	..	1	..	1	1	..	..	..	..	..	1	2	..	6		
♂7007(14) × ♀5158(5)	..	..	1	..	..	..	1	..	1	1	1	..	..	..	5	} 7.50	14.50
" × ♀5940(5)....	..	1	1	1	..	1	..	..	..	..	..	1	..	..	5		
" × ♀5601(6)....	..	..	1	..	..	..	..	..	..	..	1	1	..	..	3		
" × ♀5920(6)...	1	..	1	..	..	..	..	..	1	..	1	..	..	..	4	} 7.25	14.72
" × ♀5645(6)....	..	..	1	1	..	..	1	1	..	..	1	1	..	2	8		
" × ♀6570(6)....	..	..	..	1	..	..	..	..	1	2	1	..	..		5		
Totals..............	5	7	13	13	15	8	2	2	1	2	8	18	24	12	130	7.04	15.56

TABLE 24.—*Grade distribution of the young produced by a cross of tan Dutch with self animals heterozygous for tan.*

Parents.	Grades of young.					Totals.	Means, Dutch young.
	0	1	2	2	4		
♂7142(4) × ♀6000(0).......	6	..	1	6	..	13	2.86
" × ♀6119(0).......	16	1	5	8	1	31	2.60
" × ♀6122(0).......	2	..	2	2	..	6	2.50
" × ♀6124(0).......	2	..	..	1	3	6	3.75
" × ♀6380(0).......	3	1	2	2	..	8	2.20
" × ♀7529(0).......	5	1	..	1	3	10	3.20
" × ♀7677(0).......	7	..	1	2	2	12	3.20
" × ♀8063(0).......	3	..	..	2	..	5	3.00
Totals..................	44	3	11	24	9	91	2.83

TABLE 25.—*Grade distribution of the F₁ young produced by the cross of dark Dutch with tan Dutch.*

Parents.	Grades of young.			Totals.	Means.
	0	1	2		
♂7142(4)T × ♀7641(2)D.........	2	3	..	5	....
" × ♀6058(4)D.........	..	1	2	3	....
♂6701(5)D × ♀7209(2)T.........	..	5	..	5	....
♂5757(3)T × ♀5170(6)D.........	..	2	..	2	....
" × ♀5939(6)D.........	2	5	2	9	....
Totals........................	4	16	4	24	1.00

TABLE 26.—*Grade distribution of the F₁ young produced by the cross of white Dutch with tan Dutch.*

Parents.	Grades of young.							Totals.	Mean.
	6	7	8	9	10	11	12		
♂5757(3)T × ♀7003(16)W.....	1	6	..	1	..	..	..	8	....
" × ♀7185(16)W.....	1	3	..	..	..	..	..	4	....
♂6175(17)W × ♀6424(3)T.......	..	..	..	..	..	..	3	3	....
" × ♀6539(4)T.......	2	2	1	..	..	..	..	5	....
Totals........................	4	11	1	1	..	..	3	20	7.70

TABLE 27.—*Grade distribution of the F₁ young from crosses of dark Dutch with self and of white Dutch with self.*

Parents.	Grades of young.				Totals.	Means.
	0	1	2	3		
(Dark × self.)						
♂6701(5) × ♀7413(0)...........	3	1	..	..	4	...
" × ♀8012(0)...........	9	..	..	..	9	...
Totals........................	12	1	..	..	13	.08
(White × self.)						
♂6175(17) × ♀6133(0)..........	..	..	5	..	..	....
" × ♀7123(0)..........	..	4	2	..	..	....
" × ♀7124(0)..........	..	11	4	2	..	....
" × ♀8265(0)..........	..	3	2	..	..	....
Totals........................	..	18	13	2	33	1.51

TABLE 28.—*Grade distribution of the F_2 young from the several crosses made between the three types of Dutch and between white Dutch and self.*

Cross.	Grades of young.																		Totals.	Means.
	0	1	2	3	4	5	6	7	8	9	10	11	12	13	14	15	16	17		
White × self..............	41	45	22	14	9	5	5	4	5	5	8	4	8	5	6	4	1	..	191	.....
White × dark (tables 16–18).	..	..	1	1	2	12	15	6	5	7	3	4	..	1	1	6	13	2	56+25	5.82 and 14.40
White × tan..............	..	..	..	1	7	12	13	14	18	25	10	12	10	4	4	..	..	..	130	.....
Dark × tan..............	89	84	25	8	11	17	8	4	14	9	4	2	..	..	..	..	..	..	275	.....

TABLE 29.—*Grade distribution of young produced by other crosses of F_1 animals.*

Cross.	Grades of young.																		Totals.	Means.
	0	1	2	3	4	5	6	7	8	9	10	11	12	13	14	15	16	17		
F_1 (white × self)×white	..	2	3	4	11	11	14	9	3	1	..	..	..	1	14	24	13	6	58+58	5.15 and 15.15
F_1 (white × self)× self	65	26	3	..	..	..	..	..	..	..	..	..	..	..	..	..	..	..	94	.....
F_1 (dark × self) × dark	5	36	9	7	2	..	..	..	..	..	..	..	..	..	..	..	..	..	59	.....
F_1 (white × dark)×dark	..	3	12	11	36	33	30	12	2	3	1	..	..	..	..	..	..	..	143	.....
F_1 (white × tan)× white	..	..	..	..	..	..	..	1	2	5	8	8	6	13	3	18	7	2	73	.....
F_1 (dark × tan) × dark	4	12	6	3	..	..	..	..	..	..	..	..	..	..	..	..	..	..	25	.....
F_1 (dark × tan) × tan..	19	9	1	10	13	14	4	4	1	1	..	..	..	..	..	..	..	..	76	.....
F_1 (dark × tan) × white	1	1	5	10	23	14	5	7	..	3	..	4	12	8	10	3	5	1	112	.....

TABLE 30.—*Grade distribution of young sired by the original English male, 2545, grade $2\frac{1}{4}$.*

A. MOTHER ENGLISH.

Mother.	Grade.	Self young.	English young of grade—										Total English.
			1	$1\frac{1}{4}$	$1\frac{1}{2}$	$1\frac{3}{4}$	2	$2\frac{1}{4}$	$2\frac{1}{2}$	$2\frac{3}{4}$	3	?	
2649	2	1	..	..	..	..	1	..	1	..	1	..	3
2650	3	5	2	1	..	..	2	2	..	1	1	4	13
2651	$2\frac{3}{4}$	2	2	..	..	..	1	..	..	..	..	2	5
Totals ..	..	8	4	1	..	..	4	2	1	1	2	6	21
Means ..			1.05				2.32						

B. MOTHER SELF.

Mother.	Self young.	English young of grade—												Total English.
		$1\frac{1}{2}$	$1\frac{3}{4}$	2	$2\frac{1}{4}$	$2\frac{1}{2}$	$2\frac{3}{4}$	3	$3\frac{1}{4}$	$3\frac{1}{2}$	$3\frac{3}{4}$	4	?	
1443	2	..	..	..	..	..	..	..	..	..	..	..	3	3
1492	5	..	..	..	..	..	..	..	..	1	..	1	3	5
2053	2	..	..	..	..	..	..	..	..	..	..	..	5	5
2502	5	..	..	..	..	..	..	..	..	1	..	..	2	3
2867	3	..	..	..	..	..	..	..	..	..	..	..	3	3
2912	0	2	..	1	..	1	..	1	..	..	..	..	..	5
2916	1	..	..	..	..	..	..	..	..	..	1	..	1	2
Totals .	18	2	..	1	..	1	..	1	..	2	1	1	17	26
Mean ..		2.80												

TABLE 31.—*Grade distribution of young sired by English male 2711, grade 4, generation 1, ancestor of all English in the selection series.*

A. MOTHER ENGLISH.

Mother.	Grade.	Generation.	Self young.	¾	1	1¼	1½	1¾	2	2¼	2½	2¾	3	3¼	3½	3¾	4	4¼	4½	4¾	5	?	Total English.	Means, higher group.
5051	3½	1½	2	.	1	.	.	.	.	.	.	1	.	.	.	1	.	.	.	.	.	1	4	} 3.83
5052	3½	1½	2	1	.	.	.	.	1	1	.	.	.	.	1	.	.	.	.	.	.	.	4	
5206	4¾	2½	0	.	.	.	1	.	.	.	.	.	1	.	.	1	1	.	1	1	.	.	6	3.80
5083	5	1½	1	.	.	.	2	1	.	.	.	.	.	.	.	.	.	.	1	.	.	.	4	
5084	5	1½	2	.	.	.	1	.	.	.	.	.	.	.	1	.	.	1	.	.	.	.	3	} 4.25
5402	5	1½	1	.	.	2	.	.	.	.	.	.	.	.	.	.	.	.	.	.	.	3	5	
Totals			8	1	1	2	4	2	1	.	.	1	.	.	2	.	.	2	2	.	1	3	.	.	1	3	26	
Means							1.52						3.93										

B. MOTHER SELF.

Mother.	Self young.	3	3¼	3½	3¾	4	4¼	4½	4¾	5	Total English.
2765	1	..	2	1	..	..	..	..	..	..	3
2770	1	..	..	..	..	2	..	1	..	2	5
2840	4	..	1	1	1	..	.	..	..	..	3
2862	4	..	..	..	..	1	1	..	..	..	2
2867	3	1	1	..	..	..	1	..	..	2	5
2878	0	..	..	2	1	..	..	..	..	..	3
2947	4	..	..	..	..	..	..	..	..	..	..
2948	4	..	..	..	..	2	1	..	..	..	3
2983	4	1	1	..	..	..	..	..	..	1	3
3019	3	1	..	..	..	..	..	..	..	..	1
Totals	28	3	5	4	2	5	3	1	..	5	28
Mean						3.89					

TABLE 32.—*Grade distribution of young sired by English male 5086, grade 4½, generation 1½.*

A. MOTHER ENGLISH.

Mother.	Grade.	Generation.	Self young.	½	¾	1	1¼	1½	1¾	2	2¼	2½	2¾	3	3¼	3½	3¾	4	4¼	4½	4¾	5	Total English.
5051	3½	1½	2	.	.	1	1	1	.	.	1	.	.	.	1	1	1	1	.	.	.	1	9
5052	3½	1½	8	.	.	2	1	1	.	.	2	.	.	.	1	2	.	.	2	2	1	2	2	.	.	18
5188	4¼	2½	2	.	.	.	.	.	.	.	.	.	.	.	.	.	.	.	1	.	.	.	1
5206	4¾	2½	2	.	.	.	1	.	.	.	.	.	.	.	.	.	1	2	.	.	.	.	4
5053	5	1½	2	.	.	.	1	.	.	.	.	.	.	.	.	.	2	.	.	1	.	.	4
5083	5	1½	6	.	.	.	.	.	.	.	.	.	.	.	.	1	1	2	1	1	.	.	6
5084	5	1½	3	.	1	.	.	.	1	.	.	.	.	.	.	.	.	1	.	.	.	.	3
5101	5	1½	3	.	1	.	.	2	1	.	.	.	.	.	1	1	1	.	.	.	1	3	1	12
5102	5	1½	6	1	1	.	.	.	.	.	.	.	.	.	1	.	.	1	1	.	.	.	5
5398	5	1¾	1	.	.	.	.	.	.	.	.	.	.	.	.	.	.	1	1	.	.	.	2
Totals			35	1	3	5	4	4	.	3	.	.	.	3	4	5	8	8	4	5	6	1	64
Means								1.38								3.96						

TABLE 32, continued.

B. MOTHER SELF.

Mother.	Self young.	English young of grade—																Total English.
		1¼	1½	1¾	2	2¼	2½	2¾	3	3¼	3½	3¾	4	4¼	4½	4¾	5	
4146	3	1	..	..	1	..	..	1	..	..	2	..	1	..	..	..	1	7
4147	2	..	..	..	..	..	..	..	1	..	..	..	..	..	..	..	..	1
4148	3	..	..	..	..	..	..	..	1	..	1	..	..	..	..	..	..	2
Totals....	8	1	..	..	1	..	..	1	2	..	3	..	1	..	..	..	1	10
Mean..		3.15																

TABLE 33.—*Grade distribution of young sired by English male 5375, grade 4½, generation 2½.*

Mother.	Grade.	Generation.	Self young.	English young of grade—																		Total English.	Means.
				½	¾	1	1¼	1½	1¾	2	2¼	2½	2¾	3	3¼	3½	3¾	4	4¼	4½	4¾	5		
5051	3½	1½	2	..	1	..	..	..	..	..	1	..	1	1	..	1	..	..	1	..	..	6	} 3.40
5052	3½	1½	2	1	..	..	..	..	2	..	..	1	1	..	1	1	1	..	..	..	..	8	
5188	4¼	2½	1	..	1	..	1	..	..	..	..	..	..	1	1	..	..	..	..	..	..	4	} 3.85
5205	4¼	2½	1	..	1	1	..	..	..	..	..	..	..	..	..	..	3	..	..	..	..	5	
5206	4¾	2½	1	..	..	1	..	..	..	..	..	..	..	1	1	1	..	..	..	..	..	4	3.75
5083	5	1½	2	..	..	..	..	..	1	..	..	..	..	..	..	..	..	..	..	..	..	1	
5084	5	1½	5	..	1	..	..	..	..	..	..	..	..	..	1	3	..	..	..	..	1	6	} 4.11
5102	5	1½	9	..	3	..	..	..	..	1	..	..	..	..	2	..	..	1	..	1	8	
5398	5	1¾	2	1	1	..	..	..	..	..	..	..	..	..	2	..	..	..	..	..	..	4	
Totals..			25	2	1	7	1	2	2	2	..	1	1	2	1	3	9	8	..	2	..	2	46	
Means .				1.20								3.79												

TABLE 34.—*Grade distribution of young sired by English male 5555, grade 4¾, generation 2½.*

Mother.	Grade.	Generation.	Self young.	English young of grade—																			Total English.	Means, higher group.
				½	¾	1	1¼	1½	1¾	2	2¼	2½	2¾	3	3¼	3½	3¾	4	4¼	4½	4¾	5	?		
5557	3	2½	3	..	..	1	..	..	..	..	..	..	..	1	2	..	1	..	2	..	..	..	7	} 4.04
5051	3½	1½	3	..	..	..	..	..	..	..	..	..	1	..	1	..	5	..	1	1	1	1	11	
5052	3½	1½	2	..	..	1	..	1	..	..	..	..	..	..	..	..	..	1	..	..	..	..	3	
5561	3½	2½	1	..	..	..	..	..	..	..	..	..	..	..	..	..	..	1	..	..	..	..	1	
5701	4	3½	..	..	..	..	..	2	1	..	..	..	..	1	2	..	3	2	..	..	..	..	11	}
5752	4	3	..	..	..	..	1	..	1	..	..	..	..	..	..	..	1	1	1	..	..	..	5	
5993	4	3	1	..	..	..	..	..	..	..	..	..	..	..	..	..	1	1	..	..	..	..	2	
5188	4¼	2½	9	..	..	2	1	..	2	..	..	..	..	1	1	3	3	1	5	..	..	..	19	} 4.47
5205	4¼	2½	1	..	..	..	..	1	..	..	..	..	..	..	..	2	..	..	..	..	..	..	3	
5672	4½	3	6	..	..	2	..	..	..	..	..	..	..	1	..	..	..	..	..	1	1	..	5	
5769	4½	3½	2	..	..	..	..	..	..	1	..	..	..	1	..	1	..	..	..	..	..	..	3	
5952	4½	3	1	..	..	..	..	..	..	..	..	..	..	1	1	1	..	2	..	..	..	..	5	
6074	4½	3½	2	..	..	..	..	..	..	..	..	..	..	..	1	1	3	..	..	..	..	..	5	
5206	4¾	2½	6	..	..	2	1	1	..	1	..	..	..	2	..	4	2	1	6	2	..	..	22	
5793	4¾	3½	2	..	..	..	..	1	1	..	..	..	..	..	..	1	3	..	..	2	..	..	8	
5801	4¾	3	3	..	..	..	2	..	..	1	..	..	..	1	1	2	3	4	1	..	..	..	15	
5988	4¾	3	1	..	..	..	..	..	..	..	..	..	..	..	1	1	..	..	..	..	..	..	4	} 4.44
5084	5	1½	12	..	..	2	3	1	..	..	..	..	1	..	..	1	1	2	2	2	..	..	15	
5102	5	1½	2	1	1	..	..	1	..	1	..	..	..	1	..	..	1	..	4	..	2	..	12	
5398	5	1¾	7	..	..	1	..	1	..	..	..	..	..	..	..	..	2	2	2	1	..	..	9	
5951	5	3	1	..	..	2	..	..	..	..	..	..	..	1	..	..	..	..	..	..	..	..	3	
Totals.			65	1	1	12	9	7	6	4	..	1	..	1	2	7	7	18	16	26	29	19	2	168	
Means.				1.36								4.40													

TABLE 35.—*Grade distribution of young sired by English male 6370, grade 5, generation 3.*

Mother.	Grade.	Generation.	Self young.	English young of grade— 1	1¼	1½	1¾	2	2¼	2½	2¾	3	3¼	3½	3¾	4	4¼	4½	4¾	5	Total English.	Means, higher group.
5752	4	3	4	..	..	..	..	..	..	..	..	..	..	..	..	..	..	..	..	4	4	
5993	4	3	2	..	..	..	..	3	..	..	..	..	..	..	..	..	..	2	1	1	7	
5188	4¼	2½	0	..	..	..	1	..	..	..	..	..	..	..	..	..	1	2	2	1	7	
5672	4½	3	0	..	..	..	..	..	..	..	..	..	..	..	..	..	..	..	..	1	1	
5769	4¼	3½	0	..	..	..	..	..	..	..	..	..	..	..	..	1	3	..	1	2	7	4.67
6074	4½	3½	4	..	1	..	..	1	..	..	..	..	..	..	..	..	..	..	..	..	2	
6089	4½	3½	2	..	..	1	1	..	..	..	..	..	..	..	..	..	1	1	1	2	7	
7193	4½	4¼	0	..	..	..	..	..	1	..	..	..	..	..	..	..	..	2	1	..	4	
6815	4¾	3½	3	1	2	..	..	..	1	1	..	..	..	..	1	..	..	4	2	5	17	
6369	5	3	1	..	..	1	..	..	..	..	..	..	..	..	..	..	..	2	1	..	4	
6417	5	3	1	..	..	1	..	..	..	..	..	..	..	..	..	..	2	..	..	..	3	4.63
6693	5	4¼	1	..	..	..	..	1	..	1	..	..	..	..	..	..	1	1	2	1	7	
6788	5	4½	3	..	..	..	..	..	..	..	..	..	..	..	..	..	..	1		1	2	
Totals	......	.....	21	1	3	3	2	5	2	2	..	..	..	..	1	1	8	15	11	18	72	
Mean.	....	.....	......	1.79											4.66							

TABLE 36.—*Grade distribution of young sired by English male 6420, grade 4½, generation 3.*

Mother.	Grade.	Generation.	Self young.	English young of grade— 1	1¼	1½	1¾	2	2¼	2½	2¾	3	3¼	3½	3¾	4	4¼	4½	4¾	5	Total English.
5993	4	3	1	1		..	..	..	..	..	..	..	..	..	1	2	..	.			4
5188	4¼	2½	0	..	..	..	2	..	..	..	..	..	..	2	..	..	..	1	..	..	5
6074	4½	3½	0	..	1	..	..	..	..	..	..	..	..	..	..	3	..	..	2	1	7
5102	5	1½	1	1	1	..	..	..	1	..	1	1	..	..	..	..	..	..	..	..	5
6417	5	3	1	..	..	..	..	..	..	..	..	..	..	..	..	..	1	1	1	..	3
6841	5	4¼	1	..	..	..	..	1	..	..	..	..	..	..	..	..	..	1	2	1	5
Totals	....	....	4	2	2	..	2	1	1	..	1	1	..	2	1	5	1	3	5	2	29
Means.	....	....	....	1.80										4.33							

TABLE 37.—*Grade distribution of young sired by English male 6071, grade 5, generation 3½*

Mother.	Grade.	Generation.	Self young.	English young of grade— 1	1¼	1½	1¾	2	2¼	2½	2¾	3	3¼	3½	3¾	4	4¼	4½	4¾	5	Total English.	Means, higher group.
5051	3½	1½	1	2	..	..	..	..	..	..	..	..	..	..	3	..	..	..	..	..	5	3.75
5752	4	3	1	1	1	..	..	..	..	..	..	..	..	..	..	1	1	..	..	2	6	4.56
5769	4¼	3½	0	..	..	1	..	..	..	..	..	..	..	..	..	..	..	..	1	2	4	4.83
5891	4½	3½	2	..	1	..	..	..	..	..	..	..	..	..	1	..	1	..	..	..	3	4.25
6074	4½	3½	1	1	1	..	1	1	..	..	..	..	..	..	..	..	..	..	1	..	5	
5206	4¾	2½	0	..	..	1	..	..	..	..	..	..	..	2	..	1	..	..	..	..	4	
5793	4¾	3½	2	..	..	..	..	..	..	..	..	..	..	..	..	..	..	..	1	4	5	
5988	4¾	3	3	..	..	..	2	1	..	..	..	..	..	..	1	..	..	1	2	..	7	4.47
6073	4¾	3½	1	..	..	..	..	1	..	2	..	..	..	..	..	..	..	..	1	..	4	
6189	4¾	3	1	..	..	..	2	..	..	..	..	..	..	..	..	..	1	1	..	..	4	
5084	5	1½	2	..	..	1	1	..	..	..	..	..	..	1	..	..	..	1	..	..	4	4.12
5951	5	3	2	..	..	..	1	..	..	..	..	..	..	..	..	..	2	..	..	..	3	
Totals	....	.....	16	4	3	3	7	3	..	2	..	..	..	3	5	2	5	3	6	8	54	
Means	....	.....	....	1.61										4.39								

TABLE 38.—*Grade distribution of young sired by English male 6072, grade 5, generation 3½.*

A. MOTHER HETEROZYGOUS ENGLISH.

Mother.	Grade.	Generation.	Self young.	English young of grade— ¾	1	1¼	1½	1¾	2	2¼	2½	2¾	3	3¼	3½	3¾	4	4¼	4½	4¾	5	5¼	Total English.	Means, higher group.
5752	4	3	2	..	..	1	..	..	..	..	..	..	..	..	..	..	1	1	..	1	..	..	4	4.50
5701	4	3½	3	..	..	..	1	1	..	..	..	1	..	..	..	..	1	1	1	1	2	..	9	
5993	4	3	2	..	..	..	..	..	..	..	..	..	1	..	..	..	1	..	..	..	1	..	3	
5188	4¼	2½	0	..	..	..	1	..	..	1	1	..	..	..	..	..	..	..	1	..	1	..	5	4.42
5769	4¼	3½	4	..	1	..	..	1	1	..	..	2	..	..	..	2	1	..	1	3	..	..	12	
5891	4½	3½	3	..	..	..	1	1	..	..	..	..	..	..	..	..	..	1	..	..	..	..	3	4.58
6074	4½	3½	2	..	..	..	..	..	..	1	..	..	..	..	..	..	1	..	..	..	..	..	2	
6079	4½	4	1	..	..	..	..	..	..	..	..	..	..	..	..	..	..	..	1	1	2	..	4	
6089	4½	3½	7	..	..	..	1	..	3	..	..	..	..	..	..	..	..	1	..	1	2	..	8	
6452	4½	4	2	..	..	1	..	..	1	..	..	..	..	..	..	..	..	1	5	4	..	..	12	
6795	4½	3½	3	..	..	..	..	1	1	1	..	..	..	..	..	1	..	2	1	4	1	..	12	
5206	4¾	2½	3	..	..	..	1	..	..	..	..	..	..	..	..	..	..	..	..	..	..	..	1	4.62
5801	4¾	3	4	..	..	1	..	1	..	..	..	..	..	..	..	..	3	1	2	2	1	..	11	
5988	4¾	3	6	..	..	2	..	..	1	1	..	..	..	..	..	..	3	..	3	2	3	..	15	
6189	4¾	3	4	..	..	..	..	1	..	..	4	..	..	..	..	..	..	..	1	3	4	..	13	
6264	4¾	3¾	6	..	..	..	..	..	1	..	..	..	..	..	..	..	..	1	1	1	1	..	5	
7450	4¾	4¼	5	..	..	..	..	..	..	..	..	..	..	..	..	..	..	1	..	..	..	..	1	
7817	4¾	3½	1	..	..	..	..	..	..	1	..	..	..	..	..	..	..	..	..	..	2	..	3	
8813	4¾	4¾	0	..	..	..	..	..	..	2	..	..	..	..	..	..	..	..	..	1	1	..	4	
5084	5	1½	2	..	..	..	..	2	1	..	..	..	..	..	..	..	1	..	..	1	2	..	7	4.69
5102	5	1½	9	..	1	1	1	2	1	..	..	1	..	..	..	..	..	2	1	4	2	..	16	
6188	5	3	1	..	..	..	..	..	..	..	1	..	..	..	..	..	..	..	4	1	1	..	7	
6369	5	3	2	..	..	..	..	2	2	..	..	..	..	..	..	..	..	..	3	1	2	..	10	
6416	5	3	2	..	..	2	..	1	2	..	1	..	2	..	..	..	..	..	4	4	5	1	22	
6417	5	3	4	..	..	..	..	..	..	..	..	..	..	..	..	..	..	..	1	3	1	..	5	
6622	5	3¾	1	..	..	..	..	1	..	..	..	..	..	..	..	..	..	..	..	3	1	..	5	
6693	5	4¼	0	1	..	..	..	2	..	..	1	..	..	..	..	..	1	1	1	4	1	..	12	
6841	5	4¼	2	..	..	..	1	..	1	2	..	1	1	..	..	..	..	2	4	7	..	..	19	
7476	5	4¼	0	..	..	..	..	..	..	..	..	..	..	..	..	..	..	..	..	4	..	..	4	
Totals..	......	......	81	1	2	8	7	16	15	9	8	5	4	..	..	3	13	15	35	56	36	1	234	
Means .	......	......	......	1.97												4.63								

B. MOTHER HOMOZYGOUS ENGLISH.

Mother.	Grade.	Generation.	Self young.	English young of grade— 1	1¼	1½	1¾	2	2¼	2½	2¾	3	3¼	3½	3¾	4	4¼	4½	4¾	5	Total English.
5733	2	3	0	..	..	..	..	..	1	..	3	1	..	..	..	..	..	..	2	1	8
7535	1¼	4	0	..	..	..	..	..	..	..	..	..	..	..	1	..	..	..	..	..	1
7814	1½	3½	0	..	2	1	1	2	..	..	1	1	1	1	..	2	..	..	2	..	14
8704	1¾	4¾	0	1	..	..	..	2	..	..	..	..	..	..	1	1	..	..	..	..	5
Totals...	......	......		1	2	1	1	4	1	..	4	2	1	1	2	3	..	..	4	1	28
Means...	......	......	......	2.11										4.17							

C. MOTHER SELF.

Mother.	Self young.	English young of— 3½	3¾	4	4¼	4½	4¾	Total English.
7123	6	..	..	1	1	..	1	3
7878	3	1	..	..	..	2	1	4
Totals....	9	1	..	1	1	2	2	7
Mean....	......	4.32						

TABLE 39.—*Grade distribution of young sired by English male 6964, grade 5, generation 4½.*

Mother.	Grade.	Generation.	Self young.	English young of grade— ¾	1	1¼	1½	1¾	2	2¼	2½	2¾	3	3¼	3½	3¾	4	4¼	4½	4¾	5	Total English.	Means, higher group.
6074	4½	3½	4	1	..	1	..	..	..	..	..	..	..	..	..	..	..	..	4	2	..	8	4.64
7193	4½	4¼	2	..	..	..	..	..	1	..	..	..	..	..	..	..	..	..	1	1	1	4	
7475	4¾	4¼	1	1	1	..	..	..	..	..	..	..	..	..	..	..	1	..	..	..	..	3	
7356	4¾	4¼	2	..	..	..	..	..	..	..	..	..	..	..	..	..	..	..	..	..	2	2	4.78
7817	4¾	3½	1	..	1	..	1	..	..	..	1	..	..	..	..	..	..	..	..	3	2	8	
5102	5	1½	4	..	1	2	..	..	1	2	..	..	..	..	..	..	1	..	2	2	..	11	
6417	5	3	1	..	1	..	2	1	1	..	..	..	..	..	..	..	..	..	1	..	..	6	
6693	5	4¼	2	..	..	1	1	..	..	..	..	..	..	..	..	..	..	1	..	2	1	6	4.65
6841	5	4¼	4	..	..	..	..	..	..	..	..	..	..	..	..	..	..	..	..	..	2	2	
Totals..	....	.....	21	2	4	4	4	1	3	2	1	..	..	..	..	..	2	1	8	10	8	50	
Means.	....	.....	....	1.49													4.68						

TABLE 40.—*Grade distribution of young sired by English male 7699, grade 5, generation 4¼.*

A. MOTHER HETEROZYGOUS ENGLISH.

Mother.	Grade.	Generation.	Self young.	English young of grade— 1	1¼	1½	1¾	2	2¼	2½	2¾	3	3¼	3½	3¾	4	4¼	4½	4¾	5	5¼	?	Total English.	Means, higher group.
9449	3½	3½	2	.	..	..	1	1	..	..	..	..	..	..	..	..	..	1	..	..	..	.	3	4.50
6089	4½	3½	2	.	..	..	..	2	..	..	..	..	..	..	..	..	..	1	1	..	..	.	4	4.62
5988	4¾	3	1	.	1	..	..	1	..	..	..	..	..	..	1	.	..	1	..	1	..	3	8	
6795	4¾	3½	4	.	..	..	..	..	..	1	1	..	..	..	1	1	..	..	1	1	..	.	6	
7300	4¾	4½	0	.	..	1	..	..	..	2	..	2	..	..	..	1	..	..	2	1	1	.	10	
7356	4¾	4¼	0	.	..	..	..	..	..	..	..	..	..	..	..	.	..	..	1	1	..	1	3	4.66
7389	4¾	4¼	0	.	..	..	..	1	..	..	..	..	..	..	..	.	..	1	..	1	..	.	3	
7475	4¾	4¼	8	.	1	2	1	..	3	..	..	..	..	..	..	.	2	4	3	1	..	.	17	
7817	4¾	3½	1	.	..	..	..	1	..	1	..	..	..	..	..	.	..	1	..	3	..	.	6	
9350	4¾	5¼	2	.	1	1	2	..	1	..	1	1	..	..	1	.	1	..	1	6	1	.	17	
5084	5	1½	2	1	..	..	1	..	..	..	..	..	..	1	..	1	..	1	2	2	..	.	9	
6369	5	3	4	.	..	..	..	..	1	..	..	1	..	..	..	.	..	1	5	2	..	.	10	
6416	5	3	3	.	2	..	..	1	..	..	2	..	..	..	..	.	..	..	5	1	..	.	11	
6417	5	3	4	.	..	..	1	..	..	..	..	..	1	..	..	2	..	1	3	1	..	1	10	
6622	5	3¾	5	.	..	..	..	..	1	..	1	..	..	..	..	.	..	..	2	3	2	.	9	
6841	5	3¾	3	.	..	..	..	1	3	1	1	3	..	1	..	1	1	1	3	4	..	.	20	
7325	5	3½	0	.	..	..	..	1	..	..	..	..	..	..	..	1	..	..	..	1	..	.	3	4.80
7476	5	4¼	3	.	..	..	..	1	..	..	..	..	..	..	..	.	..	..	1	5	1	.	8	
7906	5	4	3	.	..	..	..	1	..	1	2	..	..	..	..	.	..	..	3	2	..	.	9	
8257	5	4¾	0	.	..	1	..	2	1	..	2	1	1	..	..	1	..	..	2	5	..	.	16	
8259	5	4¾	7	.	..	..	..	..	..	..	..	..	..	..	..	.	1	3	4	6	1	.	15	
9349	5	5¼	0	.	..	..	..	1	..	..	1	3	..	..	..	.	..	..	1	6	2	2	16	
9363	5	5	0	.	..	..	..	..	..	..	..	..	..	..	..	.	..	1	3	..	..	1	5	
9091	5	5¼	3	.	..	..	..	..	..	..	..	..	..	..	..	.	1	..	..	1	..	.	2	4.81
9535	5¼	3¼	7	.	..	..	..	..	..	..	..	2	..	..	..	.	..	..	..	2	..	.	4	
Totals	....	.....	64	1	5	5	6	14	10	6	11	13	2	2	3	8	6	17	43	56	8	8	224	
Means	....	.....	....	2.31												4.80								

TABLE 40, continued.

B. MOTHER HOMOZYGOUS ENGLISH OR SELF.

Mother	Grade	Generation	Self young	English young of grade—																	Total English	Means, higher group
				$1\frac{1}{4}$	$1\frac{1}{2}$	$1\frac{3}{4}$	2	$2\frac{1}{4}$	$2\frac{1}{2}$	$2\frac{3}{4}$	3	$3\frac{1}{4}$	$3\frac{1}{2}$	$3\frac{3}{4}$	4	$4\frac{1}{4}$	$4\frac{1}{2}$	$4\frac{3}{4}$	5	$5\frac{1}{4}$		
Hom. Eng., 8704	$1\frac{3}{4}$	$4\frac{3}{4}$	..	1	..	..	1	1	..	..	1	..	..	..	..	..	2	3	1	1	11	4.78
Self, 7124	..	..	3	..	..	..	..	..	..	..	..	..	..	..	..	1	..	8	6	1	16	} 4.74
" 8251	..	..	2	..	..	..	..	..	..	..	..	..	..	..	1	2	..	..	..	..	3	

TABLE 41.—*Grade distribution of young sired by English male 9532, grade 5, generation 4.*

Mother	Grade	Generation	Self young	English young of grade—																	Total English	Means, lower group	Means, higher group
				$1\frac{1}{4}$	$1\frac{1}{2}$	$1\frac{3}{4}$	2	$2\frac{1}{4}$	$2\frac{1}{2}$	$2\frac{3}{4}$	3	$3\frac{1}{4}$	$3\frac{1}{2}$	$3\frac{3}{4}$	4	$4\frac{1}{4}$	$4\frac{1}{2}$	$4\frac{3}{4}$	5	$5\frac{1}{4}$			
Het. Eng. 6795	$4\frac{3}{4}$	$3\frac{1}{2}$	3	1	..	..	1	..	..	..	..	..	..	..	..	1	2	..	..	..	5	} 2.46	4.69
" " 7475	$4\frac{3}{4}$	$4\frac{1}{4}$	1	..	..	..	..	..	..	..	..	..	..	..	..	1	3	..	..	..	4		
" " 9362	$4\frac{3}{4}$	5	2	..	..	..	..	..	2	1	..	..	1	..	..	..	1	..	..	..	5		
" " 9538	$4\frac{3}{4}$	$4\frac{1}{2}$	1	..	..	..	..	..	..	..	..	..	..	..	..	..	1	..	..	..	1		
" " 6369	5	3	1	..	..	..	1	..	..	..	..	..	1	..	..	..	..	2	..	..	4	} 2.30	4.70
" " 6622	5	$3\frac{3}{4}$	0	..	1	..	..	..	..	..	..	..	..	..	..	2	2	..	..	..	5		
" " 8257	5	$4\frac{3}{4}$	2	..	..	..	..	..	2	..	..	..	..	..	..	..	1	1	..	..	4		
" " 8259	5	$4\frac{3}{4}$	0	..	..	..	..	1	..	..	..	..	..	..	..	1	4	..	..	..	6		
" " 9535	$5\frac{1}{4}$	$3\frac{1}{2}$	0	..	..	..	..	..	..	2	..	..	..	..	..	..	1	..	1	..	4	} 2.85	4.94
" " 9592	$5\frac{1}{4}$	$4\frac{1}{4}$	0	..	..	..	1	..	..	1	1	..	..	..	..	..	1	1	..	..	5		
Totals			10	1	..	1	2	2	2	4	1	2	1	..	1	..	5	16	4	1	43	2.53	4.73
Means						2.53									4.73								
Hom. Eng. 7814	$1\frac{1}{2}$	$2\frac{1}{2}$	..	..	..	1	..	..	..	..	..	..	..	2	1	1	1	..	..	..	6	1.75	4.30
Self, 8251	..	..	3	..	..	..	..	..	..	..	..	..	..	..	..	1	2	..	..	..	3	} 4.80
" 8265	..	..	3	..	..	..	..	..	..	..	..	..	..	..	..	..	2	..	..	..	2	

TABLE 42.—*Grade distribution of young sired by English male 9806, grade 5, generation 5½.*

Mother	Grade	Generation	Self young	English young of grade—													Total English
				2	$2\frac{1}{4}$	$2\frac{1}{2}$	$2\frac{3}{4}$	3	$3\frac{1}{4}$	$3\frac{1}{2}$	$3\frac{3}{4}$	4	$4\frac{1}{4}$	$4\frac{1}{2}$	$4\frac{3}{4}$	5	
6844	$3\frac{3}{4}$	$4\frac{1}{4}$	2	..	..	..	1	..	..	..	..	..	..	1	1	2	5
9871	$4\frac{3}{4}$	4	2	..	3	..	..	..	..	..	..	..	1	..	1	4	9
6369	5	3	1	..	..	..	..	2	..	..	..	..	..	..	..	1	3
6622	5	$3\frac{3}{4}$	1	1	1	..	..	1	..	..	..	..	..	..	1	1	5
456	5	$4\frac{1}{4}$	1	..	..	1	..	..	..	..	..	..	..	..	1	2	4
Totals			7	1	4	1	1	3	..	..	..	..	1	1	4	10	26
Means						2.52							4.86				

the distribution of young sired ... English male 6001, grade 5, generation 4½.

Genera-tion num-ber.	Self gradings.	English gradings of grade—																			Total Eng-lish.	Means, higher group.
	5														4	2		6		4.64
	5												1	1	1		4		
	5	5	2							1					2		3	2	4.78
	6												3	3	8				
	5				4				1	2	2		11						
	5		2	1						1			5		4.65				
	5		1						1		2	1	6						
	6											2	2						
	50														50				

the distribution of young sires. English male 7003, grade 5, generation 4½.

4. Mothers Heterozygous Females.

Genera-tion num-ber.	Self grading.	English gradings of grades—																				Total Eng-lish.	Means, higher group.
	5													1			3		4.50
	5									1	1			4			4.62		
	5							1			1	3	3	8					
	6							1			1	6							
	5					2				1	1	10							
	5									3	3	4.66							
	6						2	4	2	17									
	5							3	5										
	5				3	4	4	1	17										
	5				5	4	9												
	5				3	2	19												
	5			5	1	11													
	5		3	5	5	1	14												
	5		2	2	9														
	5		3	3	4	20													
	5		4	5	1	3		4.80											
	5		5	1	8														
	5		5	2	16														
	5		4	4	15														
	5	1	6	2	2	16													
	5	3	3	5															

TABLE 40, continued.

B. MOTHER HOMOZYGOUS ENGLISH OR SELF.

Mother.	Grade.	Generation.	Self young	1¼	1½	2	2¼	2½	2¾	3	3¼	3½	3¾	4	4¼	4½	4¾	5	5¼	Total English.	Means, higher group.
				English young of grade—																	
Hom. Eng., 8704	1¾	4¾		1		1	1	..	..	1	..	..	..	..	..	2	3	1	1	11	4.78
Self, 7124	..	..	3					..	..	..	..	..	..	..	1	..	8	6	1	16	4.74
" 8351	..	..	2					..	..	..	..	..	..	1	2	..	..	..	..	3	

TABLE 41.—*Grade distribution of young sired by English male 9632, grade 5, generation 4.*

Mother.	Grade.	Generation.	Self young	1¼	1½	1¾	2¼	2½	2¾	3	3¼	3½	3¾	4	4¼	4½	4¾	5	5¼	Total English.	Means, lower group.	Means, higher group.
				English young of grade—																		
Het. Eng. 6795	4¾	3½					1		..	..	..	..	..	..	..	1	2	..	..	5	2.46	4.69
" " 7475	4¾	4¾	1						..	..	..	..	..	..	..	1	3	..	..	4		
" " 9362	4¾	5					..	2	1	..	..	1	..	..	..	..	1	..	..	5		
" " 9638	4¾	4½	1						..	..	..	..	..	..	..	..	1	..	..	1		
" " 6369	5	3	1						..	..	..	..	..	1	..	..	..	2	..	4	2.30	4.70
" " 6622	5	3¾							..	..	..	..	..	..	..	2	2	..	..	5		
" " 8257	5	4¾	2						2	..	..	..	..	..	..	..	1	1	..	4		
" " 8259	5	4¾					..	1	..	..	..	..	..	..	..	1	4	..	..	6		
" " 9635	5½	3¾							..	..	2	..	..	..	..	..	1	..	1	4	2.85	4.94
" " 9692	5½	4¾							1	1	..	..	..	..	..	..	1	1	..	5		
Totals..		..	10	1		2	2	2	4	1	2	1		1		5	16	4	1	43	2.53	4.73
Means......	..	..		2.53										4.73								
Hom. Eng. 7814	1½	2½	..				..	..	..	..	..	..	..	2	1	1	1	..	..	6	1.75	4.30
Self, 8251	..	..	3	..		..	..	..	..	..	..	..	..	..	..	1	2	..	..	3	.....	4.80
" 8265	..	..	3			..	..	..	..	..	..	..	..	..	..	..	..	2	..	2		

TABLE 42.—*Grade distribution of young sired by English male 9806, grade 5, generation 5½.*

Mother.	Grade.	Generation.	Self young.	2¼	2½	2¾	3	3¼	3½	3¾	4	4¼	4½	4¾	5	Total
				English young of grade—												
6844	3¾	4¼	2	..	..	1	..	..	..	..	..	..	1	1	2	
9871	4¾	4	2	3	..	..	..	..	..	..	..	1	..	1	4	
6369	5	3	1	..	..	..	2	..	..	..	..	..	..	..	1	
6622	5	3¾	1	1	..	..	1	..	..	..	..	..	..	1	1	5
456	5	4¼	1	..	1	..	..	..	..	..	..	..	..	1	2	4
Totals........					1	1	3	..	..	..	..	1	1	4	10	26
Means...					[illegible]							4.86				

TABLE 43.—*Grade distribution of young sired by English male 1212, grade 5, generation 5.*

A. MOTHER HETEROZYGOUS ENGLISH.

Mother.	Grade.	Generation.	Self young.	English young of grade— 2	2¼	2½	2¾	3	3¼	3½	3¾	4	4¼	4½	4¾	5	5¼	Total English.	Means higher group.
9943	4½	5¼	2	1	..	..	..	..	..	..	..	..	..	..	..	4	..	5	....
455	4¾	4¼	1	..	..	..	1	..	..	..	..	..	..	..	..	2	..	3	
539	4¾	5¼	...	..	..	..	..	..	..	..	1	..	..	..	1	..	2	4	
1360	4¾	5½	7	1	..	..	..	..	..	1	..	..	..	..	..	5	..	7	4.97
6795	4¾	3½	4	1	..	..	..	..	..	..	..	..	..	..	2	2	..	5	
7475	4¾	4¼	...	..	..	..	1	..	..	..	..	..	..	..	1	2	..	4	
8257	5	4¾	1	..	1	..	1	..	..	..	..	..	..	..	..	2	1	5	
452	5	5¾	2	..	..	..	..	1	..	..	..	..	..	..	..	..	..	1	
9372	5	4	1	..	..	..	..	..	..	..	..	..	..	..	1	4	..	5	
230	5	4½	...	..	..	..	..	..	..	..	..	..	..	..	2	3	..	5	
6841	5	4¼	1	..	..	..	..	..	..	..	..	..	..	..	1	3	..	4	
537	5	5¼	2	..	..	..	..	..	1	..	..	1	..	..	..	1	..	3	
6622	5	3¾	...	..	..	..	..	..	..	..	..	..	..	..	1	1	..	2	
112	5	5¾	4	2	..	..	..	2	..	..	1	..	..	..	..	2	..	7	4.97
1222	5	5¼	...	..	..	..	..	..	..	..	..	..	..	..	..	1	..	1	
596	5	5¾	2	..	..	..	..	1	..	..	..	1	..	..	..	1	..	3	
1613	5	4½	...	..	..	..	..	..	..	1	1	..	..	..	..	5	..	7	
1860	5	6½	...	..	..	..	..	..	..	..	..	..	..	..	1	2	..	3	
6369	5	3	1	..	..	..	..	..	..	..	..	..	..	..	..	..	..	..	
9363	5	5	1	..	..	..	..	..	..	..	..	..	..	..	1	1	1	3	
111	5¼	5¾	...	1	..	..	2	..	..	..	..	..	..	..	..	2	3	8	5.15
597	5¼	5¾	...	1	..	..	..	..	1	..	..	..	..	..	..	..	..	2	
Totals..	......	......	29	7	1	..	5	4	2	2	3	2	..	..	11	43	7	87	
Means...	......	......	......	2.87											4.98				

B. MOTHER HOMOZYGOUS ENGLISH.

Mother.	Grade.	Generation.	English young of grade— 1¾	2	2¼	2½	2¾	3	3¼	3½	3¾	4	4¼	4½	4¾	5	Total English.
8682	1¾	4¾	..	..	1	..	1	4	..	..	..	..	1	..	1	3	11
9953	2¾	5	..	..	..	..	..	..	..	..	..	..	..	..	1	2	3
9684	3¼	4½	..	1	..	..	..	..	..	..	..	..	..	..	2	..	3
9970	3½	4½	1	..	..	..	..	..	..	..	..	..	1	1	3	..	6
Totals...	....	....	1	1	1	..	1	4	..	..	..	..	2	1	7	5	23
Means....	....	....	2.59									4.75					

TABLE 44.—*Grade distribution of young sired by English male 534, grade 5¼, generation 5*

A. MOTHER HETEROZYGOUS ENGLISH.

Mother.	Grade.	Generation.	Self young.	2	$2\frac{1}{4}$	$2\frac{1}{2}$	$2\frac{3}{4}$	3	$3\frac{1}{4}$	$3\frac{1}{2}$	$3\frac{3}{4}$	4	$4\frac{1}{4}$	$4\frac{1}{2}$	$4\frac{3}{4}$	5	$5\frac{1}{4}$	Total English.	Means, higher group.
9684	$3\frac{1}{4}$	$4\frac{1}{2}$	2									1						1	....
9943	$4\frac{1}{2}$	$5\frac{1}{4}$	1												1	2		3	....
7475	$4\frac{3}{4}$	$4\frac{1}{4}$	2	1									1		1	1		4	
9871	$4\frac{3}{4}$	4	2								1					1		2	
360	$4\frac{3}{4}$	$5\frac{1}{4}$	..	1						1						2		4	
455	$4\frac{3}{4}$	$4\frac{1}{4}$	2							3						3		6	4.87
584	$4\frac{3}{4}$	$5\frac{1}{2}$	2	1												1		2	
1360	$4\frac{3}{4}$	$5\frac{1}{2}$	..	2												3		5	
539	$4\frac{3}{4}$	$5\frac{1}{4}$	..				1	1		1						2		5	
6369	5	3	1					2	2							1		5	
9363	5	5	1					1							1	5		7	
7906	5	4	..														2	2	
258	5	$5\frac{3}{4}$	..					1	2							2	1	6	
452	5	$5\frac{3}{4}$	2													1	2	3	5.02
230	5	$4\frac{1}{2}$	1							1						1		2	
456	5	$4\frac{1}{4}$	1					1	1							2		4	
537	5	$5\frac{1}{4}$	..												1			1	
1222	5	$5\frac{1}{4}$	1													2		2	
6841	5	$4\frac{1}{4}$	..													2		2	
112	5	$5\frac{3}{4}$	1				1								1	3		5	
9091	5	$5\frac{1}{4}$	1						1							3	1	5	
2325	$5\frac{1}{4}$	$5\frac{1}{2}$	..							2	1					3		6	5.06
111	$5\frac{1}{4}$	$5\frac{3}{4}$	..													2	2	4	
597	$5\frac{1}{4}$	$5\frac{1}{2}$	1				1		1							1		3	
9592	$5\frac{1}{4}$	$4\frac{1}{4}$	..								1					3	1	5	
Totals ...			21	5			6	11	3	4	3	1	1		9	42	9	94	
Means.							2.95							4.97					

B. MOTHER HOMOZYGOUS ENGLISH.

Mother.	Grade.	Generation.	$1\frac{1}{4}$	$1\frac{1}{2}$	$1\frac{3}{4}$	2	$2\frac{1}{4}$	$2\frac{1}{2}$	$2\frac{3}{4}$	3	$3\frac{1}{4}$	$3\frac{1}{2}$	$3\frac{3}{4}$	4	$4\frac{1}{4}$	$4\frac{1}{2}$	$4\frac{3}{4}$	5	Total English.
7814	$1\frac{1}{2}$	$3\frac{1}{2}$	1		1	1										1	1	1	6
8682	$1\frac{3}{4}$	$4\frac{3}{4}$					1									3	1		5
9953	$2\frac{3}{4}$	5							2									1	3
Totals...			1		1	1	1		2							4	2	2	14
Means..					2.37											4.69			

TABLE 45.—*Grade distribution of young sired by homozygous English male 1173, grade 3¼, generation 5.*

A. MOTHER HETEROZYGOUS ENGLISH.

Mother.	Grade.	Generation.	English young of grade—																Totals.
			1½	1¾	2	2¼	2½	2¾	3	3¼	3½	3¾	4	4¼	4½	4¾	5	5¼	
9684	3¼	4½	..	1	..	1	..	1	1	..	..	..	..	..	1	..	..	..	5
9970	3½	4½	1	3	2	..	..	..	..	..	..	..	2	..	..	..	..	..	8
7300	4¾	4½	..	..	1	2	..	1	..	2	2	..	..	..	..	..	..	..	8
455	4¾	4¼	..	..	..	..	..	..	..	..	..	..	..	..	..	3	1	..	5
584	4¾	5½	..	..	2	1	..	..	..	..	..	..	..	..	1	1	3	..	8
6369	5	3	..	..	..	..	2	1	..	..	..	..	..	..	..	..	1	1	5
8257	5	4¾	..	..	..	..	..	..	..	..	..	..	..	..	..	3	..	..	3
1222	5	5¼	..	..	..	..	..	..	1	..	..	..	..	..	..	..	3	..	4
Totals ...	....	....	1	4	5	4	2	4	2	2	2	..	2	..	2	7	8	1	46
Means ...	....	....	2.42										4.77						

B. MOTHER HOMOZYGOUS ENGLISH.

Mother.	Grade.	Generation.	English young of grade—					Totals.
			2	2¼	2½	2¾	3	
7814	1½	3½	1	2	1	..	..	4
9953	2¾	5	..	..	..	2	4	6
Totals ..	....	....	1	2	1	2	4	10
Mean....	....	....	2.65					

C. MOTHER SELF.

Mother.	English young.					Totals.
	4¼	4½	4¾	5	5¼	
7123	..	2	4	2	..	8
7124	1	2	3	9	1	16
8251	..	..	2	..	..	2
9318	..	..	..	5	..	5
Totals	1	4	9	16	1	31
Mean....	4.84					

TABLE 46.—*Summary of young of selected heterozygous English males by heterozygous English females.*

Sire.	Grade.	Generation.	Self young.	Homozygous English young.	Mean grade.	Heterozygous English young.	Mean grade.	Ungraded English young.	Total English	Total young.
2711	4	1	8	12	1.52	11	3.93	3	26	34
5086	4½	1½	35	20	1.38	44	3.96	..	64	99
5375	4½	2½	25	17	1.20	29	3.79	..	46	71
5555	4¾	2½	65	41	1.36	125	4.40	2	168	233
6370	5	3	21	18	1.79	54	4.66	..	72	93
6420	4½	3	4	10	1.80	19	4.33	..	29	33
6071	5	3½	16	22	1.61	32	4.39	..	54	70
6072	5	3½	81	75	1.97	159	4.63	1	235	316
6964	5	4½	21	21	1.49	29	4.68	..	50	71
7699	5	4¼	64	75	2.31	149	4.80	8	232	296
9532	5	4¼	10	16	2.53	27	4.73	..	43	53
9806	5	5½	7	10	2.52	16	4.86	..	26	33
1212	5	5	29	26	2.87	61	4.98	..	87	116
534	5¼	5	21	32	2.95	62	4.97	..	94	115
Totals..	..	..	407	395	...	817	...	14	1,226	1,633
Percent.	..	..	24.9	24.2	...	50.0	...	0.9	75.1	.....

TABLE 47.—*Line of advance in the selection experiment with heterozygous English males.*

Sire.	No. of homozygous young.	Mean of homozygous young.	No. of heterozygous young.	Mean of heterozygous young.	Advance, homozygous young.	Advance, heterozygous young.
2545...............	5	1.05	9	2.80	....	....
2711...............	12	1.52	11	3.93	.47	1.13
5555...............	41	1.36	125	4.40	—.16	.47
6072...............	81	1.97	159	4.63	.61	.23
7699...............	75	2.31	149	4.80	.34	.17
1212 and 534.......	58	2.91	123	4.97	.60	.17
Total advance	..	....	....	....	1.86	2.17
Average advance ..	..	....	....	....	.37	.43

TABLE 48.—*Relative frequency of cross-overs among the gametes formed by F₁ rats of the two sexes.*

Series.	F₁ parent.	Dark-eyed young.	Red-eyed or pink-eyed yellow young.	Total young.	Percentage dark-eyed young.	Percent cross-over gametes.
Repulsion.....	Female....	101	837	938	10.8	21.5
"	Male......	73	703	776	9.4	18.8
Coupling	Female....	699	1,046	1,745	40.0	19.8
"	Male......	556	731	1,287	43.2	13.5

BIBLIOGRAPHY.

CASTLE, W. E., and P. B. HADLEY.
  1915. The English rabbit and the question of Mendelian unit-character constancy. Proc. Nat. Ac. Sci. 1, pp. 39–42, 6 figs.

—— and JOHN C. PHILLIPS.
  1914. Piebald rats and selection. Carnegie Inst. Wash. Pub. No. 195, 56 pp., 3 pl.

—— and S. WRIGHT.
  1916. Studies of inheritance in guinea-pigs and rats. Carnegie Inst. Wash. Pub. 241, 192 pp., 7 pl.

HALDANE, J. B. S., A. D. SPRUNT, and N. M. HALDANE, 1915. Reduplication in mice. Jour. Genet. 5, 133–135. [Reference from Detlefsen (1918) Genetics, 3, p. 597.]

KING, HELEN D.
  1918. The effects of inbreeding on the fertility and on the constitutional vigor of the albino rat. Jour. Exp. Zool. 26, pp. 335–378.

PUNNETT, R. C.
  1912. Inheritance of coat-colour in rabbits. Jour. Genet., 2, pp. 221–238, 3 pl.

PLATE 1

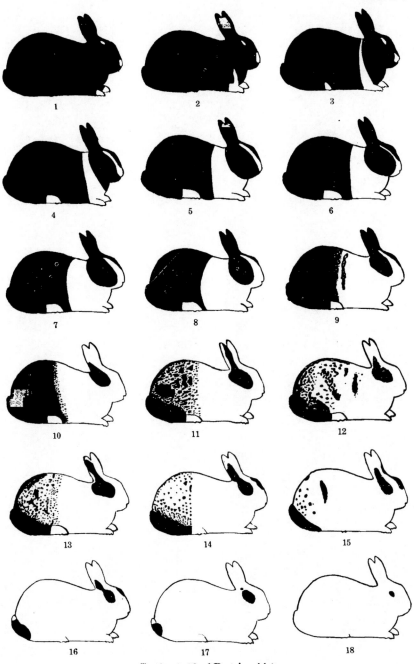

Grades 1–18 of Dutch rabbits.

PLATE 2

Fig. 19, a "white" Dutch rabbit, ♀7934, grade 15   Fig. 20, a "dark" Dutch rabbit, ♂6701, grade 5.   Fig. 21, a "tan" Dutch rabbit, ♀747, grade 3

PLATE 3

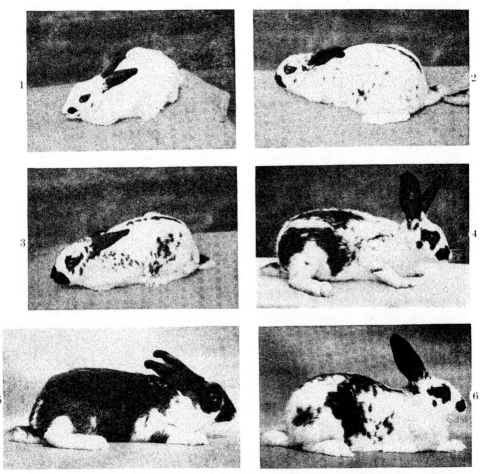

FIGS 1-5. Photographs of English rabbits adopted as grades 1-5 in classifying the variations observed in the English pattern. Those shown in figs. 1 and 2 were homozygous, those shown in figs 3 to 5 were heterozygous.

FIG. 6 A "high-grade" homozygous English rabbit, ♂1173, grade 3½, genetically comparable with a grade 5 heterozygous English. Compare fig 5.